Ultimate FCAT*

Cynthia and Drew Johnson

*Florida Comprehensive Assessment Test

Simon & Schuster

New York · London · Singapore · Sydney · Toronto

Kaplan Books
Published by Simon & Schuster
1230 Avenue of the Americas
New York, NY 10020

For bulk sales to schools, please contact: Order Department, Simon & Schuster, 100 Front Street, Riverside, NJ 08075. Phone: 1-800-223-2336. Fax: 1-800-943-9831.

Project Editor: Ruth Baygell
Contributing Editors: Marc Bernstein, Marcy Bullmaster
Cover Design: Cheung Tai
Production Manager: Michael Shevlin
Interior Page Design and Layout: Laurel Douglas
Production Editor: Maude Spekes
Managing Editor: Dave Chipps
Executive Editor: Del Franz

Manufactured in the United States of America

September 2000
10 9 8 7 6 5 4 3 2 1

Library of Congress Cataloging-in-Publication Data is available.

ISBN: 0-7432-0176-0

All of the practice questions in this book were created by the authors to illustrate question types. They are not actual test questions. For information on the FCAT, visit the Florida Department of Education Web site at www.firn.edu/doe/.

TABLE OF CONTENTS

ABOUT THE AUTHORS

Cynthia and Drew Johnson

Cynthia Johnson is the author of several educational books for young people, two of which received the prestigious Parent's Choice Gold Award in 1995, and were listed in *Curriculum Administrator* magazine's "Top 100" educational products for 1996. Drew Johnson is an education writer and editor, creating workbook, textbook, and Web-based education materials for children of all ages. The Johnsons have authored *Kaplan Learning Power*, a guide for improving study skills, and Kaplan's No-Stress Guides and Parent's Guides to various statewide standardized tests.

INTRODUCTION

WHY TAKING THE GRADE 10 FCAT IS LIKE DRIVING A CAR FOR THE FIRST TIME

You sit down, trying to remember all the things you know you're supposed to remember. The people around you seem to be feeling the same way you do—a little nervous. An adult looks at you and points at his watch, telling you it's time to start. There's no chance of going back now: It's time to take the Grade 10 FCAT.

Driving down the FCAT Road can sometimes be a confusing experience.

You're probably familiar with the Florida Comprehensive Assessment Test (FCAT) tests already. The version you'll take in grade 10 is similar to its grade 8 sibling: both tests cover three basic subject areas—reading, mathematics, and writing—and both are composed of multiple-choice questions, gridded-response questions, and performance tasks. Although the two tests are similar in format, the grade 10 FCAT has a much greater impact on your academic career, because Florida law now makes passing the FCAT one of the requirements for high school graduation. Starting with the class of 2002, all Florida high school students will have to consider the following mathematical equations:

You + good FCAT scores = diploma

You – good FCAT scores = retaking the grade 10 FCAT until you achieve a passing score on all subjects.

On each subject test, you'll receive an **achievement-level score**, from *level 5* (highest) to *level 1* (lowest). A score of level 2 (limited success with state standards) or level 1 (little success with state standards) means you'll have to retake that subject test.

You don't need to be Albert Einstein to recognize that the first equation is more desirable than the second. Fortunately, if you fail only one subject test—for example, if you receive a level 2 score on Math but pass the other subject tests with a Level 3 score—you'll have to take only that one subject test over again. Of course, since most people aren't thrilled about having to take *any* standardized test, this could well be called "finding the silver lining in the cloud." Yet there's no need to feel bad if you do happen to get a low score, because high FCAT scores aren't very common. In the first two years the FCAT tests were given, less than one-half of all students passed the Math test, and only about one-third passed Reading.

The FCAT was administered for the first time in 1998, and Florida made history the following year when it passed Governor Jeb Bush's A+ Education plan, initiating the first statewide voucher program in the United States. Under the plan, each of Florida's 2,500 schools receives a letter grade of A, B, C, D, or F depending on how well its students fared on the FCAT Reading, Writing, and Math tests. If a school receives an F twice in four years, then students at that school may use an education voucher, worth about $4,000, at another school of their choice. That new school can be a private school, a religious school, or a qualified public school (a public school with a C grade or higher). Two elementary schools in Pensacola became the first schools to fail twice. The constitutionality of the A+ Plan is currently being challenged in court.

How the FCAT Tests Were Born

Currently, there are two different stories circulating about the origin of the FCAT tests.

Version A:

Ten years ago, on the coldest night of winter, a 100-pound meteor crashed into a pumpkin field outside of Pensacola, Florida. The meteor cracked open on impact, and at its core were the very first FCAT tests.

Version B:

In 1993 the state of Florida started work on creating a new set of tougher school curricula. Known as the *Sunshine State Standards*, this new set of educational standards was completed in 1997. The FCAT was created to test those standards, and to see how well students were learning these subjects.

While on a stroll, an educator watches a batch of FCAT tests fall to Earth.

While most people prefer Version B because of its greater adherence to the facts, Version A does have the advantage of making a better made-for-television story. Whichever story you prefer, the truth is that the FCAT tests are here to stay, and that's why this book was written. If taking the grade 10 FCAT is like driving a car for the first time, then *Ultimate FCAT* provides you with cool wheels, a great road map, the best insurance, and all the training you need to make the experience a successful one. We've included all the techniques and strategies you need to do well on each subject test, and we've also provided FCAT-like sample questions to help you practice these skills. If you study these strategies and apply them on the test, then taking the grade 10 FCAT should change from a difficult experience into something as easy as driving to the beach on the first day of summer break.

Chapter 1

TEST-TAKING STRATEGIES

HOW TEST-TAKING STRATEGIES ARE LIKE A SWISS ARMY KNIFE

This chapter will provide you with the test-taking strategies you'll need in order to succeed on the FCAT. You might even be using some of these techniques already, but there's a big difference between using test-taking strategies *some* of the time and using them *all* of the time, for every subject test.

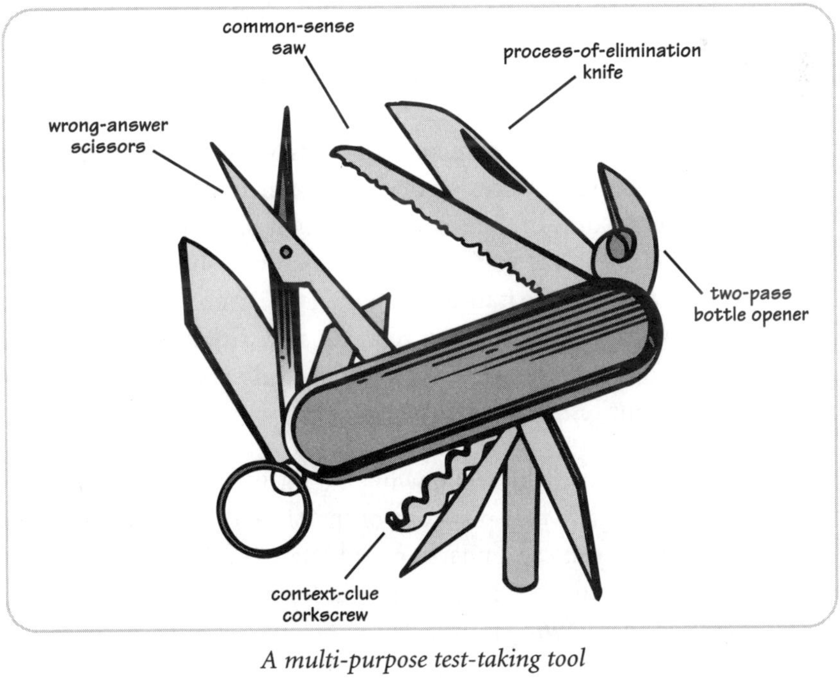

A multi-purpose test-taking tool

In this respect test-taking strategies are like that Swiss Army knife you got for Christmas: If you look at it just once and then put it in a drawer, it will never help you. If you occasionally carry it around with you, you might find some use for it, but it's more likely you'll forget it's there and not use it when a situation arises. However, if you take the knife and examine how all the different parts can be used, then you'll find you have the perfect tool for many different occasions.

This same premise holds true for the following test-taking strategies. If you take the time and practice using all of these techniques, you'll help yourself out on each MCAS subject test, and on other standardized tests you'll take in the future.

Strategy 1

Know the format of each FCAT subject exam ***before*** test day.

Make sure you approach the FCAT knowing how the test is set up; for example, how many multiple-choice questions and performance tasks are on each subject test, and how much each question is worth. This is important for two reasons. First, you need to know what to expect in order to prepare for it—you don't want to mentally prepare for a killer essay only to find out that the questions are all multiple-choice. Furthermore, knowing the format beforehand helps eliminate some of the anxiety you might be feeling about the test. If you know what questions you'll be facing, then you should feel empowered by that knowledge, since it means that the FCAT won't surprise or upset you. However, if the grade 10 FCAT remains some mysterious exam you have to pass in order to graduate, you'll probably feel nervous about it—and that could affect your test performance.

In terms of the Swiss Army knife, knowing the format of a test is like using the little magnifying glass that pops out of the far end. It lets you examine the FCAT up close so you can understand it better. The format of each subject test will be discussed later in the book, but for now you should familiarize yourself with the following four facts.

The Big Four Format Facts

1. FCAT tests are given in "sessions."

The Writing FCAT consists of a single 45-minute session, while the Math and Reading FCATs consist of two 80-minute sessions each. Most school districts give the Math and Reading FCATs over a four-day period, with one 80-minute session per day.

2. There are three different question types on the FCAT.

Multiple-choice questions—worth 1 point each, you choose the best answer out of four choices.

Short- and long-answer questions *(or performance tasks)*—short-answer questions are worth 0–2 points each, and the long-answer questions are worth 0–4 points each. For these, you'll have to write out an answer in your own words. Your answer must include an *explanation*, because you will be scored on your work as well as on your final response. Performance tasks are called "Think, Solve, and Explain" questions in math and "Read, Think, and Explain" questions in reading.

Gridded-response questions—worth 1 point each, these appear only on the Math test. Without having any choices to pick from, you'll solve a math problem and then fill in a response grid with your numerical answer. These are worth only 1 point because they're much simpler than open-response problems.

3. *The question breakdown on each subject test is fairly consistent from year to year.*

Keep in mind that the *exact* mixture of questions varies on each test administration, but for the most part, it remains the same.

FCAT Subject Test Format

Reading:	50 multiple-choice questions 6–8 short-answer questions 2–3 long-answer questions
Mathematics:	45 multiple-choice and gridded-response questions 6–8 short-answer questions 2–3 long-answer questions
Writing:	1 essay

Note: These are the tests you'll take in February, and they'll be used to determine whether you'll graduate high school. There are also two additional FCAT tests given in March, in Math and Reading Comprehension. These multiple-choice-only exams are used only to evaluate your performance in relation to that of students throughout the country.

4. *Short- and long-answer questions can be identified by their icons.*

The test is designed to "flag" the questions for which you'll have to write out your thoughts. So keep an eye out for the following icons:

short-answer performance task icon (reading)

long-answer performance task icon (reading)

As you examine the icons, note that the *long-answer icon has blank lines underneath the banner*, while the short-answer icon does not. Make sure you understand the difference, because it wouldn't do you much good to spend 15 minutes on a question that's worth only 2 points. Just remember: *More lines equals longer answer.*

Strategy 2

Don't spend too much time and energy on any one question at the expense of the other questions.

On most of the tests you take in high school, you try to get the highest grade possible, which means answering every question no matter how long it takes. This isn't a good idea on the FCAT, because here, there are only two real grades: Pass or Fail. Spending too much time on a question means you've wasted precious minutes and increased your mental fatigue greatly—for a question that's worth only about $\frac{1}{70}$ of your score.

An older Florida student berates himself for taking too long on a short-answer question.

Strategy 3

Maintain a consistent pace throughout the test. Don't rush through—or spend too long on—any one question.

Proper pacing is something that all good test takers do naturally, but that can be learned quite easily by everyone else. Here are all the ingredients you'll need.

Mom's Old-Fashioned Pacing Recipe

Ingredients:

1. Two eyes (one eye will do as well).

2. A watch.
   A digital watch or stopwatch is preferable, since it provides a more accurate display. Sundials and grandfather clocks are not a good choice.

Directions:

Use ingredient 1 to look at ingredient 2. Repeat often so that you have an accurate idea of how many questions are left, and how much remaining time you have to work on them.

Granted, pacing sounds simple, but in the heat of the moment this is one strategy people often abandon first—and usually with negative results. Make sure you practice pacing before test day. In fact, you'll find that it will also help you in other parts of your life, as effective time management is a useful skill to have.

At the start of the test, glance through to see how many multiple-choice, gridded-response, and performance tasks there are. Then, set a pace that will allow you to finish the session with enough time to proofread your work.

Note: The official FCAT test directions recommend spending 10–15 minutes on each long-answer question. If there are two of these questions, you could spend 30 minutes—or one-third of your time—solving two difficult problems worth 8 points at most. For students shooting for a perfect score, this might be a good trade-off, but for most people, this isn't the most efficient use of time. *Spend more time making sure you've done the easier 1–point problems correctly.*

Pacing Rule-of-Thumb:

Multiple-choice and gridded response questions	**1–3 minutes each**
Short-answer questions	**5–7 minutes each**
Long-answer questions	**8–10 minutes each**

Consider this: If you have a test with 23 multiple-choice, 3 short-answer, and 2 long-answer questions, you'll know that you should spend about 50 minutes on multiple-choice, 10–12 minutes on short-answer, and about 16 minutes on long-answer. This will leave you with time at the end to look over your work. This last step is always crucial since it gives you a chance to catch any mistakes you've made.

By setting—and keeping—a proper pace for yourself, and by going into the test knowing what you'll see, you'll be taking control of the FCAT. In other words, go in with a game plan for how much time you'll want to spend on each question. This will be much better than just showing up and scribbling frantically once the exam begins.

Now that you know how much time you're going to spend on each question, the next step is deciding what order to do the questions in.

Strategy 4

Do all the easy, multiple-choice questions first before you tackle the more involved performance tasks.

Think, for a moment, about what the best "order" might be to work through the test. On many tests you've taken, multiple-choice questions appear early on—before the more involved questions—so you're probably accustomed to doing the easy questions first. But it's also logical to assume that you would want to get the easy questions out of the way before you attempt the more involved questions. So on the FCAT, if the performance tasks appear first, don't just jump in and work on them first, if there are multiple-choice or gridded-response questions that follow. Why spend time on a hard question when there are easier questions, and easier points, waiting for you after it?

Strategy 5

Use a ***two-pass system*** on the test. On the first pass answer the easy questions and skip the rest. Then go back a second time and answer the more challenging questions.

The best way to approach the test is to use a *two-pass system*: on the first pass answer all the easy questions. Then go through the test again a second time, and answer the more challenging questions. If you come to a multiple-choice problem that stumps you, after a minute or so just move on. You can always come back to it, and there's no reason to get bogged down when there are easier points awaiting you. On the first pass, you want to concentrate your efforts on answering the questions you know, while on the second pass you want to attack the more difficult problems. If you're working at a consistent pace, you should spend more time on the second pass to work on the more involved questions.

A farmer considers how to use the two-pass system on his crops.

Let's say you're using the two-pass system and you come up on a tough multiple-choice problem. You look at it briefly on the first pass, but the question is very unfamiliar to you, so you move on to the rest of the multiple-choice questions. On the second pass, you reread the problem, but it still doesn't make sense. Should you just skip it and move on? The answer is "Absolutely not!" *There's no guessing penalty on the FCAT,* so any question you leave blank is a missed opportunity for free points.

Strategy 6

Don't leave any questions blank. You won't lose points for wrong answers so even if you have to guess, answer every question on the test.

Some tests, like the SAT, deduct any wrong answers you make from your total score. But the FCAT does no such thing, so even if you have an impossibly hard question, you still have a one-in-four chance of getting the question right.

1. What is the proper motion of Barnard's star?*

   A. 14 millibreems

   B. 10.25 arc seconds a year

   C. 6.12 star miles

   D. The Treaty of Ghent

If you do get an incredibly difficult question like the one above right, it won't matter that it was guess; all that matters is that the proper oval has been filled in.

The "Answer every question" maxim holds true for the multiple-choice questions and for the open-ended performance tasks. Since short- and long-answer questions are worth 0–2 and 0–4 points respectively, you can earn *partial credit*. A perfect long answer is worth 4 points, but you can earn 1–3 points if you provide a partially correct response.

Imagine that after reading two consecutive reading passages you were faced with the following performance task.

2. The main characters in each story share many similarities. List three of them.

What if, even after referring to the reading passages, you can come up with only two similarities. Write them down! Perhaps you can think of only one similarity but there's another you're not sure about. Write them both down! If that second similarity turns out to be wrong, you simply won't receive credit for it. In other words, if you don't write it down, there'll be no chance to earn any points! Performance tasks are designed to make you put your

* The answer is B. This question is much tougher than an actual FCAT problem. It is used to illustrate how random guessing can get you the right answer even on an incredibly difficult problem.

thoughts down in writing, and if you're reluctant to do so, your FCAT score will suffer. You have nothing to lose and everything to gain by giving every question your best shot. In fact, your score can only go up.

Strategy 7

Use the Process of Elimination to help you find the correct answer. Try ruling out clearly wrong answers one by one.

If test-taking strategies are like a Swiss Army knife, then the Process of Elimination is the biggest, strongest knife in the set. POE is something most students know about, but rarely use; they tend to rely instead on their knowledge of the subject to answer multiple-choice questions. Knowing the answer to a question outright is great, but when that fails, POE is the most effective tool you can use to get the right answer anyway. On every multiple-choice question, *the answer is in front of you.* All that matters is whether or not you pick the correct answer. Look at the following two questions.

3. Who was the thirteenth president of the United States?
   Place your answer here. ________________

4. Who was the thirteenth president of the United States?
   A. Abraham Lincoln
   B. Leo Tolstoy
   C. Millard Fillmore
   D. Chuckles the Clown

In question 3, you either know the answer or you don't. But in question 4, you can use POE and eliminate any answer choices you think are wrong. D is out, and if there was ever a president named Tolstoy, you would have remembered it, so B must be out as well. That leaves A or C—at this point, it's a fifty-fifty shot, so take a guess and move on. You won't guess the right answer all the time, of course, but if you use POE regularly and effectively, you'll get enough right to boost your score.

Common sense is an important part of POE. In question 4, common sense told you that America would never elect a clown to be a president (although some U.S. historians would tell you that Warren G. Harding comes pretty close). Look at how common sense can help you eliminate some answer choices in the following math problem.

5. Feinberg Fabrics uses the following price list for its cotton sheets.

Area	Price
3 cm^2	$6.00
6 cm^2	$9.30
9 cm^2	$12.90
12 cm^2	$16.80

Based on this information, what would an 18 cm^2 sheet likely cost?

A. $12.00

B. $16.60

C. $18.60

D. $25.50

Before you start working out the math, use common sense and POE. According to the chart, if a 12 cm^2 cotton sheet costs $16.80, will a larger sheet (18 cm^2) cost more or less than that? A larger sheet will cost more, so the correct answer will have to be *more than* $16.80, the price of the 12 cm^2 sheet. Using POE, then, you can eliminate A and B, since those answer choices are less than $16.80. This leaves C and D, and at this point, you can take an educated guess. Look at the pricing chart, and you might note how the price appears to jump by 3 or 4 dollars each time. Answer choice C is only about 2 dollars greater than $16.80, so D would be a safer bet. And it's the right answer!

POE can be used on all the FCAT subject tests. Look back to question 1, the Barnard's star question. If you had to eliminate one answer choice, which one would it be? You might note that choice D, The Treaty of Ghent, does not appear to have much to do with stars or motion. D can therefore be crossed out, leaving you with a one-in-three chance of guessing correctly.

Besides common sense, there are other strategies for eliminating incorrect answer choices.

Strategy 8

If an answer choice looks too good to be true, it probably is. Remember that the test is designed to challenge you, so don't just jump at the first answer choice that you suspect is correct.

Typically with standardized tests, you would rush through in an attempt to answer all the questions. And many students do rush through these tests, quickly picking answer choices that "look" right. Even if you see an answer that "looks" right in 15 seconds, try to resist the urge to select it and move on. Remember, these tests are designed to be challenging for you, so few questions will be as clear-cut as they may seem. As stated earlier in the pacing suggestions, spend 1–3 minutes on each multiple-choice question. Let's look at question 4 again.

Look back at question 1. Does any answer choice look too good to be true? Since the word *star* appears in the question as well as in answer choice C, it's unlikely that C is correct—that would be too easy! And as it turns out, the answer is B.

4. Who was the thirteenth president of the United States?
   A. Abraham Lincoln
   B. Leo Tolstoy
   C. Millard Fillmore
   D. Chuckles the Clown

Of these four names, almost everybody will recognize one president: Abraham Lincoln. If you were in a hurry, this question would take 15 seconds, since you would likely pick the first (and perhaps only) name you recognized. But if you understand how the FCAT works, and how best to approach a standardized test, you won't make that mistake. Using POE, you could eliminate B and D, as you did earlier. That leaves A and C, and frankly, choice A, Abraham Lincoln, is just to good to be true. That leaves C, so Millard Fillmore must be the 13th president of the United States.

Since multiple-choice questions comprise about 60 percent of the FCAT, using POE consistently can be a great boon to your score. Throughout this book, be sure your test your POE skills on every multiple-choice question you work on.

Strategy 9

Be of sound (and well-rested) mind and body for the test.

All the studying and practice you do for the FCAT could be lost if you show up on test day cranky, tired, and hungry. Getting a good night's sleep and having a healthy breakfast are essential for a good test performance. It's hard to concentrate on a test when your stomach is grumbling so loud it sounds like Mozart's *Requiem Mass* is being performed in your duodenum. Make sure you eat a good breakfast, but don't eat so heavily that you find yourself dying for a nap. A bowl of cereal, toast, and fruit is a good way to start the day, and if your school allows it, bring along an easy-to-eat snack.

In addition to a nutritious meal, a good night's rest is vital to having a clear head on test day. This means no cramming! Staying up late into the night studying for the test is not a good idea, since that will tire you out and is bound to increase your stress level.

Do your studying ahead of time. That way, your confidence will be highest right before the test. If you do want to review some things on the day before the test, go over more general items, such as the test format and general test-taking strategies. This will help you more than cramming on subject matter.

A positive attitude on test day is worth more than any single fact you might study the night before the exam.

Instead of studying on the night before the test, you're better off doing something that relaxes you and takes your mind off the exam. Don't exhaust yourself physically—no extreme sports! Play a board game with family or friends or watch a video. If you like, a short walk or a quick hike in a nearby park is also a good idea. But remember, if you do go on a hike, be sure to bring along your Swiss Army knife and use it in as many ways as possible.

Strategy Review

Strategy 1: Know the format of each FCAT subject exam before test day.

Strategy 2: Don't spend too much time on any one question at the expense of the other questions.

Strategy 3: Maintain a consistent pace throughout the test. Don't rush through–or spend too long on–any one question.

Strategy 4: Do all the easy multiple-choice questions first before you tackle the more involved performance tasks.

Strategy 5: Use a two-pass system on the test. On the first pass answer the easy questions. Then go through the test a second time and answer the more challenging questions.

Strategy 6: Don't leave any questions blank, even if you have to guess.

Strategy 7: Use the Process of Elimination (POE) to help you find the correct answer.

Strategy 8: If an answer choice looks too good to be true, it probably is.

Strategy 9: Be of sound (and well-rested) mind and body on test day. A positive attitude on test day is worth more than any single fact you might study the night before an FCAT session.

Chapter 2

READING

HOW ANSWERING FCAT READING QUESTIONS IS LIKE COMING UP WITH SEVEN SYNONYMS FOR *SQUEAL*

From memory, can you come up with seven words that mean the same thing as *squeal*? *Cry* is a good choice, and so are *scream* and *yell*. *Shriek* makes four synonyms, and after that . . . well, there's always *grumpy*, *sneezy*, and *bashful*. Wait—those last three are Seven Dwarves, not seven synonyms.

Coming up with seven definitions for *squeal* at the drop of a hat isn't an easy task. But what if you were asked the same question, and this time you could use a thesaurus to help you? Since a thesaurus is a reference book devoted to synonyms, the task becomes much easier: Just thumb through it until you reach *squeal*, and look at all the synonyms. You'll find *cry*, *scream*, *yell*, *shriek*, *wail*, *yelp*, and *whine*, not to mention *shrill*, *screech*, *peep*, *bawl*, and *squeak*.

It's obviously easier to refer to the thesaurus than to use just your memory. The same holds true for the Reading FCAT. Even so, many students will approach these reading passages in the least effective way; they'll read a passage and then attempt to answer the questions from memory. It's like struggling to come up with seven synonyms for *squeal* when you have a thesaurus sitting in front of you. On this section of the FCAT, you need to treat all the reading passages like the thesaurus, and continuously refer to them.

Strategy

The answer to every multiple-choice question lies somewhere in the reading passage, and that's always where you should look.

Looking back at a reading passage for an answer might seem a waste of time; you could argue "Why should I look back? The correct answer is right in front of me in the form of four answer choices!" And while that's true, it's also true that three of those four options are incorrect, and they're not all "clearly wrong." Most wrong answer choices are designed to trip you up if you don't consider them carefully, so it's always worth your while to refer to the text.

Trying to answer FCAT Reading questions without referring to the passage can lead you into a big hole.

Reading FCAT Facts

The Reading FCAT consists of two 80-minute sessions. In each session, you'll read a passage and then answer questions about it. There will be approximately 6–7 passages in all. Passages on recent tests have been in the following formats:

- nonfiction excerpt
- fiction excerpt
- excerpt from a play
- an essay
- two poems
- newspaper article
- a speech
- transcript of an interview

Of course, there's no guarantee that the test you take won't have six consecutive editorials, but since there's a great deal of consistency from year to year, you can get a good idea of what to expect.

While the types of reading passages tend to be consistent, the passage lengths are not. The average reading passage is about 900 words, but some poems will have fewer than 100 words and some essays will have over 1,500 words. Because of this large discrepancy, it's difficult to establish guidelines for pacing beforehand. You can't say, "I'll spend 5 minutes on each reading passage" since that would leave you rushing through the 1,000-word passage and wasting your time on the 100-word poem. There are no hard-and-fast pacing rules. Having said that, there are certain things you should keep in mind.

You might also see an odd type of reading passage, such as a page from an instruction manual or a poem within a larger story. Approach it as you would a regular passage, and don't let the unusual format rattle you!

Strategy

Read to understand, not to memorize, each passage. The goal of the session is to answer the questions, not to read the passages.

As you read through a text, read with an eye toward overall comprehension, not toward memorization. It's more important to understand the main idea and overall development of a passage than it is to place details into your long-term memory. Get the main idea, picking up on things like tone and the author's point of view. Then, focus in on where supporting storylines/details/facts are located—for instance, paragraph 2 deals with snakes, while paragraph 3 is about ladders. Then, head to the questions. That way, if question 1 is a main idea question, you'll be in good shape answering it, and if question 2 asks about ladders, you'll know where to look—in paragraph 3, since it deals with ladders.

By reading for the main idea/text development first and the details second, you'll be able to head right to the section of text that you need to answer a question. To help you do this, jot down notes and your own personal reminders as you read the passage for the first time. The notes don't need to be intricate, precise sentences relating exactly what occurs in each paragraph, but they should act as markers to remind you about what's in each part of the passage. These markers should help you answer questions faster, and answering questions is what the FCAT is all about.

Multiple-Choice Questions vs. Performance Tasks

Of the two types of questions on the Reading FCAT, the performance tasks (that is, the short- and long-answer questions) would unquestionably be considered harder than the multiple-choice questions—not to mention the fact that they're worth a lot more. You might think, therefore, that you should spend a lot of time on them and less time on the multiple-choice questions, which are worth only 1 point and easier to answer since you're given all four answer choices. And, following this rule of logic, you might want to rush through the multiple-choice questions so that you'll have ample time to work on the tougher performance tasks.

The problem with this line of thinking is that you'll be rushing over the easy questions that account for the bulk of your score in order to get to the harder, more time-consuming questions that account for only one-third of your score. It makes a lot more strategic sense to methodically and carefully consider all the multiple-choice questions before you work on the performance tasks.

Strategy

Take your time and carefully answer all the multiple-choice questions before you start on the open-ended performance tasks.

A hypothetical FCAT Reading test contains 44 multiple-choice questions, 7 short-answer questions, and 3 long-answer questions. A perfect score on this test would be 70 points: 44 multiple-choice points + 14 short-answer points + 12 long-answer points. While the exact question breakdown on your test may vary, it's safe to assume that:

- The performance tasks will account for one-third of your total score
- The multiple-choice problems will account for two-thirds of your total score

Because the multiple-choice questions count for most of your score, don't rush through them just to spend a lot of time on tougher questions that count for far less. If you rush on multiple-choice questions, you're likely to make careless mistakes, meaning you'd lose points needlessly.

It shouldn't be too hard to save the performance tasks for last, since they're often the final questions for each passage. In fact, not every reading passage is even followed by performance tasks. In the end, the FCAT is designed so that you can answer the questions in any order you'd like, so if you'd rather start with the final reading passage because it has no performance tasks. That's okay. If you want to start with the poetry reading passage because you're a big fan of poetry, that's fine too. You can answer the questions in whatever you want, although we still recommend you use a strategic approach and do the multiple-choice questions before tackling any performance tasks.

Strategy

Remember that you are in charge of the test, not the other way around.

Certain types of questions appear regularly on the Reading FCAT, and knowing about them ahead of time will help you on test day. Before reviewing them, however, you should read the following sample passage. As you read, remember that your goal is to understand the passage, not to memorize it.

The following excerpt has been adapted from Act I, Scene 2 from The Duchess of Malfi *by John Webster, published in 1623.*

Delio: Then the law to the duke
Is like a foul black cobweb to a spider,
He makes it his dwelling and a prison
To entangle those shall feed him.

Antonio: Most true:
He never pays debts unless they be shrewd turns,
And those he will confess that he doth owe.
Last, for his brother there, the cardinal,
They that do flatter him most say oracles
Hang at his lips; and truly I believe them,
For the darkness speaks in them.
But for their sister, the right noble duchess,
You never fix'd your eye on three fair medals,
Cast in one figure, of so different temper.
For her discourse, it is so full of rapture,
You only will begin then to be sorry
When she does end her speech, and wish, in wonder,
She held it less vain-glory, to talk much,
Than your penance to hear her: while she speaks,
She throws upon a man so sweet a look,
That it were able to raise one to a dance
That lay in a dead palsy.
. . .

Delio: Fie, Antonio,
You play yourself out with her commendation.

Antonio: I'll case the picture up, only thus much,
All her particular worth grows to this sum;
She stains the time past: lights the time to come.

The Big Four and the Little Three Question Types

If the FCAT were an *un*-standardized test, a book like this would be impossible to write. One year the FCAT might be all essays, the next year it might consist of just fill-in-the-blank questions, and the year after that it might be nothing more than a teacher standing in front of you asking, "Guess what number I'm thinking of now."

Fortunately, the FCAT is a standardized test, so the question types remain fairly consistent from year to year. There are four main types of multiple-choice question on the Reading test, and since multiple-choice questions count for two-thirds of your score, we'll call these the *Big Four*. There aren't as many performance tasks, but there are three favorite question types, so these are the *Little Three*. The more you familiarize yourself with these question types and their corresponding strategies, the easier you'll find it to answer the reading questions. And, for those questions that don't fall neatly into one of these categories (of which there are bound to be some), there's a good chance you'll be able to draw from that same strategy pool.

Don't be afraid to apply the test-taking strategies you have learned to an unfamiliar question type. If nothing else, there's always good old-fashioned Process of Elimination (POE).

Process of Elimination on the Reading Passages

The reading passages on the Reading FCAT have been chosen or written by educators. Since this is the case, you'll never see an essay that talks about how to win at gambling or a poem about how robbing banks is fun. Instead, you'll read passages that are informative, educational, and often strive to make a positive point. With this in mind, when faced with a multiple-choice question, eliminate those answer choices that are too extreme or negative. Instead, look for choices that are bland, somewhat positive, and easier to prove.

Strategy

As a general rule, when using POE on the reading questions, answer choices that are too extreme or negative are unlikely to be correct. Look for an answer that is bland, somewhat positive, and easier to prove.

1. What concept about Julius and Garcon is most thoroughly developed in the passage?
   A. Both men came from wealthy families.
   B. The two men had nothing at all in common.
   C. The two men represented different cultures.
   D. Julius and Garcon hated each other.

Even though you know nothing about the passage, you can cross out answer choices that are too extreme or negative. What would you eliminate? Choice B is rather extreme; if the two men had just *one* thing in common, it's incorrect, so it can go. D is pretty harsh, and while there's a slim chance it's true, your best bet is to cross it out. That leaves A and C, which gives you a fifty-fifty shot for guessing, and you never even saw the passage.

2. What is the best inference that can be made from the Governor's statement?
   A. There was some confusion after the hurricane.
   B. People were scared to go the disaster shelters.
   C. Most people were robbing stores downtown.
   D. Absolute panic had broken out throughout the city.

Here, C isn't a good choice because it uses the word *most*—this implies a majority of people. In order for that to be correct, the passage would have had to state that 51 percent or more of the people were robbing, and this seems extreme and too negative for the FCAT. Choice D has the same problem, with the word *absolute*. What if some people were only half-panicked? There goes your absolute. D also isn't very positive. That leaves A and B.

Choice A might work because it has the bland word *some* in it. Some can mean a little, or it could mean a lot, but it's fairly vague and therefore easy to prove. If just one person is confused, that means there's *some* confusion. As for B, a dose of common sense will cast this answer choice into a bad light. Why would people be afraid to seek a disaster shelter after a hurricane? That makes little sense. Of course, there's a chance that B is correct, but A is by far your best bet using POE. And guess what? It's the correct answer.

The Big Four Multiple-Choice Question Types

As you analyze the following question types, keep in mind the following question: "Are there any answer chances that can be eliminated using POE?" If you practice POE until it becomes second nature, your test performance is sure to improve. If you simply read about it now, but never use it again, it will be of minimal benefit.

1. Vocabulary Questions

The first category is the vocabulary question. These questions ask, in some form or another, "What does ________ mean?" To answer these correctly, you'll need to either:

A. know the definition of the word already; or

B. figure out what the word means from the surrounding text.

Option A is just nifty, but Option B is a test-taking skill that's useful to know, so that's what we'll focus on. Option B, also called *learning in context*, is something everyone can do. To illustrate this, look at the question below.

3. Joan said, "Please get the *sargfrommanhemmar* from the fridge and serve some to our guests, who are no doubt very hungry." What does *sargfrommanhemmar* mean?
   A. It is a type of food.
   B. It is a highly venomous sea snake.
   C. It is a type of drink.
   D. It is a statue made out of clay.

Based on Joan's sentence, which answer would you choose? If you choose A, you would be right. (If you choose B, you probably don't have many guests still living.) Using the words around *sargfrommanhemmar*, you can deduce that it's a type of food, which is why it was in the "fridge." Choice C is a close second guess, but why would you give a drink to people who are "hungry"? If you replace the word hungry with *thirsty* in the original sentence, C becomes a better answer.

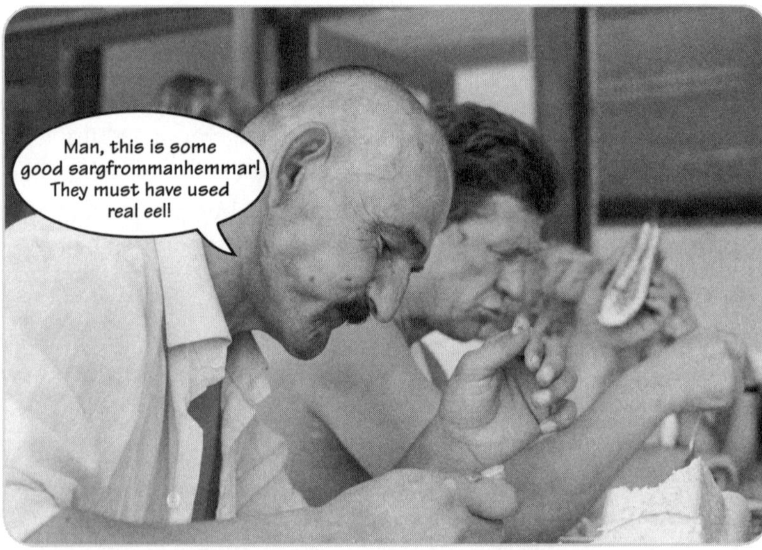

After a day of testing, the locals relax and enjoy a bowl of sargfrommanhemmar at Newt's Backway Eatery.

Strategy

On Vocabulary Questions, read the sentences around the unknown word to figure out its meaning.

Try the following question, making sure to check back to the *Duchess of Malfi* passage.

4. In line 22, what does the word *palsy* mean?
   A. eager to dance
   B. prepared to fight
   C. ready to flee
   D. unable to move

To answer this question, don't read simply the line that has *palsy* in it; always start two or three lines earlier and read through to the sentence that follows. Since you have already read through the passage for its main idea, you hopefully have the idea that "the Duchess is a really beautiful, wonderful woman," or something along those lines. With this as your framework, looking over lines 20–22 you should realize that she is so beautiful and kind, she could make someone dance who was . . . *unable to move*, choice D.

Another approach here would be to use Elimination. Some wrong answer choices are meant to mislead you, so be careful not to rush through seemingly easy questions. For example, in question 4, choice A has the word *dance* in it; since the line just before line 22 has the word *dance* in it, you might incorrectly be tempted to choose it. Cross it out instead. Then, using common sense, you might notice that line 22 actually says *dead palsy*, and this makes B and C unlikely answer choices because how could somebody be *dead* and *prepared to fight* or *flee* at the same time? You can be *dead* and *unable to move*, however.

2. Main Idea/Passage Formation Questions

These questions test your understanding of a passage's main idea. Sometimes they're straightforward, but often times, they're not. The Reading FCAT has typically included questions that ask how specific parts of a passage fit into the entire reading.

5. Which of the following statements best describes the passage?
   A. The Duchess is a good person, but her brothers are not.
   B. The Duke and the Cardinal are completely evil.
   C. Most people love the Duchess more than the Duke.
   D. The Duke rarely pays his debts.

6. Delio gives a description of the Duke, while Antonio gives a description of
   A. the Duchess only.
   B. the Duchess and the Cardinal.
   C. the Duke, Duchess, and the Cardinal.
   D. the Duke only.

Question 5 is a straightforward question about the main idea of the entire reading, while question 6 focuses on the main idea of just Antonio's monologue. In other words, it tests your understanding of a specific part of the reading.

Strategy

On Main Idea/Passage Formation Questions, look for answer choices that state the big picture and eliminate answer choices that are too small in scope.

On reading passages that have more than one paragraph, a main idea question might ask "In what paragraph did __________ begin?" This type of question is intended to see if you understand the flow of events.

To understand this strategy, look at question 5: Compare choices A and D. Both statements are true, though D is simply one small fact. Choice A is a broader statement that covers the entire reading. Specific facts from the passage are often included as misleading answer choices on main idea questions. These facts are usually accurate—they just don't represent the main idea of the text. Choice B is too extreme, since "completely evil" is impossible to prove—what if the cardinal liked to help lost puppies, and the Duke was devoted to preserving the rainforest? That would mean they're not completely evil, although from what is read, they are pretty bad.

Choice C in question 5 represents another type of wrong answer choice commonly seen on main idea questions: the probably true statement. From what you can infer from the text, chances are that you think the duchess is a likeable person, but where is that stated in the passage? The answer: Nowhere. This is why referring to the passage is always a good idea. If you suspected choice C was right, you'd have to be able to back it up—but nowhere do the characters mention that the citizens of Malfi love the Duchess more than the Duke. So while there's a good chance that that's true, you cannot safely conclude that.

Question 6 might be very easy if you had jotted down notes to yourself as you read the passage. You might have written "Duke = bad," "Cardinal = evil," and "Duchess = kind, pretty" in the margins of the text. If you did, then you're in good shape but if not, it's just a matter of referring to the passage to see who Antonio talks about. This would be a wise move, because if you don't, you might only remember what Delio said about the Duke and then pick B. In fact, Delio talks more about the Duke than Antonio, but Antonio mentions the Duke, the Cardinal and the Duchess. The question asks who the characters have described, not how much is described. The correct answer is C.

3. Inference Questions

A broad question category, Inference Questions are questions whose answers are not stated directly in the passage. Instead, the answer must be inferred from the text. For instance, although the *Duchess of Malfi* passage never exactly states the Cardinal is evil, you can infer this from the phrase "the darkness speaks in them."

In some ways, Inference Questions are like Vocabulary Questions, for which you must find clues that will help you define an unknown word. For Inference Questions, you also need to find clues that will help you draw conclusions and then answer the question.

7. According to Antonio, how did most people react when the Duchess stopped talking?
   A. They were disappointed.
   B. They were angry.
   C. They wanted to dance.
   D. They were struck by love.

Using POE, you can eliminate some choices before you refer to the passage for the right answer. Since you know that the Duchess was a good person from your first read-through, B is an unlikely choice. Choice C takes the word *dance* directly from the passage, but from a reference Antonio makes about the Duchess's *looks*, not her *speech*. About her speech, Antonio states "You only will begin, then, to be sorry/When she doth end her speech." This is the clue that should lead you to choice A, *disappointed*, which comes closest to that meaning. Again, since the answer is not actually spelled out in the passage, you'll have to read between the lines and select the answer choice that comes closest.

Strategy

On Inference Questions, find the clues in the passage that will help you determine the answer. Remember, since it won't be spelled out in the text, you'll have to infer the answer.

4. Emotion Questions

Rounding out the Big Four multiple-choice questions are Emotion Questions, which ask you to figure out the emotional state of either the author or one of the characters in the passage. In a way, these are a subset of Inference Questions, since you similarly have to infer something from clues in the text. However, since the FCAT is written by educators, a good fallback position on these questions is to look for a broad and positive answer, and then pick it.

Strategy

On many questions that ask about emotions, the answer is unlikely to be extreme or very negative. Look instead for a broad answer that communicates something positive and educational.

8. What emotion does Antonio appear to feel about the Duchess?
   A. friendship
   B. hatred
   C. mistrust
   D. admiration

9. Which word best describes the character of the Duke?
   A. nurturing
   B. petty
   C. devious
   D. pious

Question 8 shows how you can sometimes get an answer without even looking at a passage, while question 9 shows that it always helps to check your answers by looking back at the reading. For question 8, choices B and C, *hatred* and *mistrust*, are both negative, so they're likely to be incorrect. That leaves A and D, and while it's a bit of a toss-up in terms of positivity, D is the correct answer.

As useful as the "choose a broad and positive answer choice" strategy is, you can't apply it in every case. On question 9, if you hadn't read the passage you might want to guess between A and D, both positive attributes. But since the Duke is a bad man, having been compared to a "spider in foul, black cobweb," that won't work here. The answer is B or C, and from the spider reference you'll hopefully infer that *devious* is a better choice than merely *petty.*

Those are the Big Four question types. Study the strategies for them (including POE), and you should do well on them, as well as on other multiple-choice questions. Once you've answered all the multiple-choice questions, all that remains are the performance tasks.

The Little Three Performance Task Question Types

If you don't mind returning to a previous reading passage, then the best way to approach each of these Reading FCAT sessions is to answer all the multiple-choice questions, and then go back through the entire test and answer all the short- and long-answer performance tasks. The advantage of doing this is that it allows you concentrate on your multiple-choice test-taking strategies—POE, how to answer inference questions, and so on—while working on the multiple-choice questions, and then you switch mental gears and answer all the performance tasks in the appropriate manner.

Using this method, you might have to return to a reading passage, but you only have to switch test-taking techniques once. This is helpful, since it's better to tackle a test one question type at a time. However, if you're uncomfortable with this approach, it's okay to work one passage at a time, answering the multiple-choice questions first and then the related performance tasks, if there are any. Some students prefer this second approach, but keep this important fact in mind: Whichever method you choose, *do not lose track of time* on the short- or long-answer questions to the extent that you have to rush through the multiple-choice questions. This is precisely the

Pacing Performance Tasks:

Short-answer questions	**5–7 minutes each**
Long-answer questions	**8–10 minutes each**

sort of pacing trap that can hurt your score, as you sacrifice time on a hard question at the expense of simpler questions.

Performance tasks are actually a lot like the Writing FCAT essay, which will be discussed in chapter 4. For both, you're given a writing prompt, and it's up to you to present the most convincing, detailed, grammatically correct response that you can. While essays skills are discussed in detail later, the strategy that will help you the most on the short- and long-answer questions is to feel free to draw conclusions about the text as long as you can support your ideas with specific references.

Strategy

Don't be afraid to analyze the text and come to conclusions so long as you can back up your ideas.

10. Did the final paragraph provide an effective conclusion to the article? Cite evidence to support your answer.

For question 10, it makes no difference whether you say, "Yes, the final paragraph was an effective ending," or "No, it was a poor ending." All that matters is how well you back up your point of view with relevant information.

The more detailed your answer is, the better chance you have of getting the maximum number of points, but don't bust your brain trying to craft the perfect response. If you can average about 2 points for all your performance task answers—meaning that some short-answer questions would receive only 1 point, but some long-answers would get 3 or 4 points—you'll get a satisfactory grade on the Reading FCAT (provided you don't totally blow out on multiple-choice questions). Do your best to write a good response in a limited amount of time, but don't waste any precious time trying to turn a very good essay into a perfect essay. In other words, it's not worth the extra effort it would take you to turn a 3-point essay into a 4-point essay. You're better off using that time to check for errors in your multiple-choice questions first, and then in your performance task answers.

1. Two Reading Passages = One Combined Performance Task

For the most part, the Reading FCAT presents one reading passage with subsequent questions. You may see, however, two consecutive reading passages, followed by a performance task. When that happens, 99% of the time the question will ask you to compare or contrast the two passages in some manner.

Strategy

If you come across two consecutive reading passages, it's important to use information from both passages in your response. Draw up a quick comparison chart, jot down notes, and then formulate an answer.

The moment you come across two consecutive reading passages, an alarm should go off in your head signaling, "Compare/Contrast performance task a-coming!" As you read through the passages, if you come across any obvious similarities or differences, be sure to jot down a note to that effect. If you do this, when it comes time to answer the performance task, all you'll have to do is review your notes and begin composing an answer.

2. In-Your-Own-Words Performance Tasks

These questions seek your test your understanding of a passage by asking you to restate the author's meaning in your words. On the *Duchess of Malfi* passage, the question could look like the following:

> 11. In your own words, describe what Antonio thought about the Duke and the Duchess. Use specific details from the passage to support your answer.

The first part of question 11 deals with your general understanding of the passage, and you should already have that established. In other words, by the time you come to this question you should know that Antonio thinks the Duchess is awesome while brother Duke is a lowlife—if you didn't recognize this, how did you get through the multiple-choice questions? The second part of the question calls for you to back up your main point (in this case, that the Duchess is good while the Duke is a rat-fink) with specific facts from the passage. And don't just *list* the facts—explain them in your words whenever possible.

Main Point: Duchess good, Duke bad.

Specific points to back this up:

1) The Duchess looks so beautiful, she can make a paralyzed man get up and boogie (lines 20–22).
2) The Duke never pays people back if he can get away with it (lines 6 and 7).

In the limited time you have on this question, write down as many relevant details as you can that will help prove your point.

3. Understand-the-Author Performance Task

These questions boil down to whether or not you understand the author's main point. Since a good test taker will have read a passage looking for the main idea in the first place, these questions can be fairly simple to answer. For the most part, they're asking you for two things:

- What is the author's main point?
- What details make you think that?

To answer this type of performance task, state what you believe to be the author's main point, and then follow up by citing relevant parts of the passage.

When you finish the last performance task on the test, use any remaining time to check over your answers. First, make sure you have correctly answered the multiple-choice questions, and then briefly check that you bubbled in all the correct ovals. After that, review your performance task answers to correct any obvious errors.

Once you do that, you've completed that session of the Reading FCAT. To celebrate, wait until the test is over, and then let out a squeal of delight. Or if you prefer, a *scream* of delight. Or a *yell, shriek, wail, yelp, whine, screech* . . .

Chapter 3

MATHEMATICS

HOW DOING FCAT MATH IS LIKE WATCHING *BAYWATCH* IN NEPAL

Landlocked between India and Tibet and home to nine of the world's ten largest mountains, Nepal doesn't get a lot of sun 'n' surf. This could explain why many Nepalese find *Baywatch* such an interesting program, as it takes place in an area of the world that's very sunny and definitely non-Himalayan.

If you were to hang out in Kathmandu, Nepal's capital, you might catch an episode or two of the show. At first glance, the Nepali *Baywatch* would probably confuse you, since the characters would be speaking a language you're unfamiliar with. Yet once you got over your initial confusion, you would see that the show really isn't hard to follow—even in Nepalese. A lifeguard's job is to save drowning people, and this is true regardless of what language is being spoken. Watching *Baywatch* in Nepal is just like watching *Baywatch* in Florida—the characters still act the same way… if you call what they do "acting."

When you take the Math FCAT, you might come across a question so confusing that it looks like it's Nepalese in origin. If you let it rattle your nerves, chances are you'll lose confidence in your ability to solve it. But if you let your initial puzzlement pass, you should be able to figure out what the question is really asking. This is still a math test, and a fractions question is a fractions question, regardless of how it appears. Many students can't get past their anxiety in these situations, and their scores inevitably suffer. If you go into the test knowing that a few questions will appear Nepalese to you, you will be able to refocus your thoughts with confidence and figure out what's being asked of you.

Strategy

Some questions may look strange, but if you take the time to figure out what they're really asking, then doing well on them should be as easy as spending a day at the beach.

Math FCAT Breakdown

Like its Reading FCAT sibling, the Math FCAT is given in two 80-minute sessions. Here's an approximate breakdown of questions. The exact number of questions varies sightly every year, but this will give you a good idea of the question mix, and how important each question type is to your overall score.

Question Breakdown on the Math FCAT (for both sessions):

- combination of 45 multiple-choice and gridded-response questions
- 6–8 short-answer questions
- 2–3 long-answer questions

The Math test is made up of *multiple-choice questions*, for which you select from four answer choices, and *performance tasks*, for which you write out a *short* or a *long answer*. There's also a third type of question, the *gridded-response* problem, that's found only on the Math FCAT. Worth 1 point, the gridded-response is little more than a multiple-choice question without the multiple choices. You'll have to write your answer on a grid. The following chart details approximately how much these questions are worth, as well as pacing suggestions for each.

Calculators are permitted on the test, and you will be provided with a Math Reference Sheet that includes various formulas and conversions.

Question Type	Percentage of Math Score	Pacing Suggestions
multiple-choice and gridded response (1 point)	66%	1–3 minutes each
short-answer (0–2 points)	17%	5–7 minutes each
long-answer (0–4 points)	17%	8–10 minutes each

* *Percentage of Math Score* has been estimated using a hypothetical Math FCAT with 46 multiple-choice, 6 short-answer, and 3 long-answer questions. A perfect score would be 70 points.

Scoring

Multiple-choice questions and *gridded-response* questions are worth 1 point each. Performance Tasks (identified on the Math FCAT as "Think, Solve, and Explain" questions) are scored in the following way: *Short-answer* questions are worth 0–2 points. *Long-answer* questions are worth 0–4 points. You can receive partial credit for your answers, even if they aren't 100% correct. If you get half of an answer correct, you'll get half the points. Two graders will read and score your answers, basing their scores on a scoring rubric. For both performance tasks, a score of 0 means that your answer is completely incorrect or uninterpretable. A perfect score means that your response is accurate, clear and complete—though it may contain minor flaws that don't detract from your answer.

Gridded-response questions can be recognized by the grids that accompany them (see below). Though they're impervious to POE—curses!—there are still ways to improve your chances on them. Gridding in a whole number, like 86, is fairly easy. You have the option of placing the number on either *the left or the right side* of the grid.

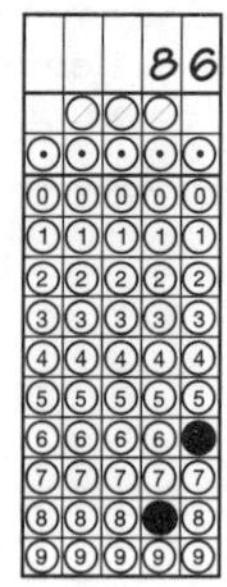

Gridding in a fraction or a decimal is a little trickier. You can use the *diagonal slash* to make a fraction, or the *decimal point* to make a decimal. So if your answer is $\frac{4}{9}$, you can write either $\frac{4}{9}$, $\frac{44}{99}$, or 0.444.

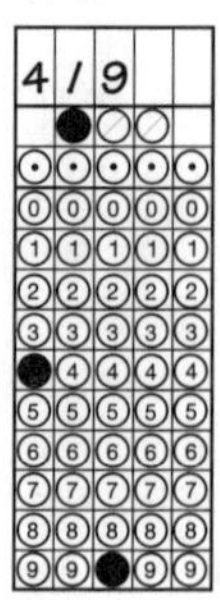

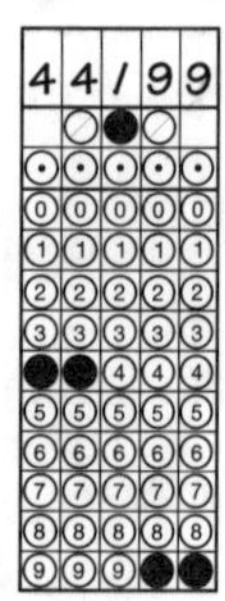

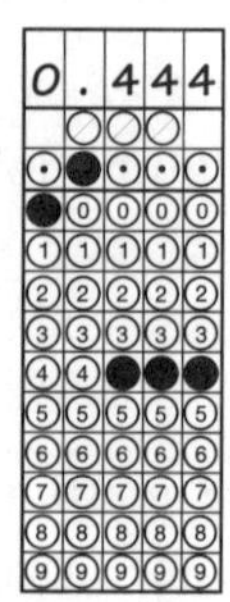

Why does $\frac{44}{99}$ work? Well, it's the same fraction as $\frac{4}{9}$, much like $\frac{3}{6}$ and $\frac{1}{2}$ mean exactly the same thing. As for the decimal 0.444, you might realize that the correct decimal is 0.444444444444444 and so on, with the four's repeating infinitely. But since the grid doesn't go off into infinity, the FDOE will accept 0.444, or even 0.44 in some instances. This might be a little confusing at first, but you can learn something from it.

Strategy

For many gridded-response questions, you'll have to express the answer as a fraction or decimal. Since fractions and decimals are more difficult to grid in properly, this is a way for the test makers to challenge you.

The test makers know that gridding in a fraction such as $\frac{45}{97}$ isn't as easy as gridding in a whole number, so many of these questions have fractions or decimals as answers. In other words, part of the challenge associated with grid-response questions is just filling them in correctly.

On the above fraction, $\frac{45}{97}$, if you were to plug that into your calculator, dividing 45 by 97 in order to find the answer as a decimal, you would get 0.4639175 before you run out of space on the calculator's LCD display. To be safe, grid in as much of this answer as possible, 0.464, making sure to round the 3 up to a 4 since there is a 9 in the ten-thousandths place. However, the FDOE has indicated that it would accept 0.46 as an answer as well, even though this shortens the number.

* On gridded-response questions, if you have to go to decimals, take care to round off your answer choice correctly.
* You *cannot* use mixed fractions on gridded- response questions.

There's one final rule about gridding in answers, and it concerns mixed fractions. A *mixed fraction* has both a whole number and a fraction, for example, $32\frac{2}{5}$ and $5\frac{1}{4}$. The best way to change these numbers into usable answers is to divide them on your calculator. That way, you'll get a decimal, which will be easier to grid. For instance, with $32\frac{2}{5}$, you would divide 2 by 5 and get 0.4—the answer to grid in would be 32.4. With $5\frac{1}{4}$, the decimal version is 5.25.

Gridded-response questions might be considered of *medium* difficulty. They're usually harder than multiple-choice questions—if only because there are no answer choices provided—but they're easier than the open-ended performance tasks. Yet since they're worth only 1 point each, work on them as you would the multiple-choice questions—*before* you attempt the short- and long-answer questions. Those questions are by nature more challenging, so you'll want to leave them for last.

Strategy

Be strategic! Answer the questions in order of how progressively difficult they are: Do the easiest questions first (multiple-choice) and then the slightly harder questions (gridded-response). Leave the most difficult questions (short- and long-answer) for last.

Since the multiple-choice questions account for the bulk of your score, doing well on them should be your top priority. After you finish those, work on the gridded-response questions, since they're only slightly harder—but be careful not to spend more than 3 minutes on each. If you want to spend 10 minutes solving a question, make sure it's a question worth 0–4 points (long-answer) and not one worth only 1 point (gridded-response).

I Calculate, Therefore I Am

As you'll be allowed to use calculators on the test, you might be thinking that you'll never have to add or subtract numbers in your head again. Hooray! 4 + 9 = . . . wait, let me get my calculator.

While it is true that a calculator can help you prevent careless errors, it won't necessarily prevent you from coming up with a wrong solution. Look at the problem below.

$4 + 9 \times 5 =$

If you're careless, you might punch 4 + 9 into your calculator, get 13, and then multiply 13 by 5 to get 65. Good for you, only this answer is wrong. You have to multiply 9 × 5 *first*, and then add 4. The correct answer is 49.

One way to prevent careless math errors like the one above is to write out all of your calculations—*before* you start using the calculator.

Strategy

Write all of your work down. Always.

Writing down all of your work is useful for many reasons. First and foremost, it gives you something to look at when you review your answers. If you write nothing down, and merely punch in numbers, there's no way to catch any mistakes you've made.

Another reason to write everything down is *partial credit*. You don't need to score a full 4 points on every long-answer question. These are the most involved, most difficult problems on the test, and if you earn 2 or 3 points on each, you'll have done a commendable job. The only way to get partial credit, though, is if the grader can look over your work and see that you were on the right track, or at least close to it. But in order for this to happen, you'll have to write something down!

One final point about the calculator. Since you'll have one, many of the math questions have been designed in such a way so that they couldn't easily be answered without one. But if you can write out the computations beforehand, be sure to do so, because this will make it easier to check your work.

Be prepared to use your calculator frequently throughout the session.

1. Which of the following numbers is smallest?
   A. $\frac{840}{48}$
   B. 8.4×101
   C. 4.8% of 84
   D. $\sqrt{84}$

Basically, question 1 tests your knowledge of math lingo and your ability to punch numbers into your calculator. Choices A and D are fairly simple computations, but B and C might be a little unfamiliar. If that's the case, just cross out D (since it's more than A and therefore can't be the correct answer) and take a guess. The correct choice is C. $4.8\% \times 84 = \frac{4.8}{100} \times 84 = 0.048 \times 84 = 4.032$, and 4.032 is the smallest number.

Why You Don't Need to Memorize $\frac{4}{3}\pi r^3$

While you may have already spent long hours memorizing formulas like the one above (which is the formula for the volume of a sphere), such diligence isn't required for the test. This is because each test booklet comes with a Reference Sheet that lists a load of formulas and conversions, free for you to use throughout the exam. These formulas include the area of a triangle, the volume of a sphere, and conversions from liters to centimeters.

Two of the most important items on the sheet are listed below. It's highly likely that you'll need to use these items on more than one Math FCAT question.

- The Pythagorean theorem, $a^2 + b^2 = c^2$, which determines the length of the third side of a right triangle if you know the other two sides; and
- The statement that 3.14 or $\frac{22}{7}$ should be used for pi.

Strategy

Be prepared to freely use the Reference Sheet on the test.

Let's examine how the formulas on the sheet can be useful.

2. If a square pyramid has a volume of 896 cubic feet, and its length and width are 8 feet each, what is its height?

On a question like this, you're sure to find both your calculator and the Reference Sheet very useful. First, write out the formula for the volume of a square pyramid from the Reference Sheet. It is: $V = \frac{1}{3}\, l \times w \times h$. Then substitute the numbers that you know:

$$896 = \frac{1}{3} \times 8 \times 8\, h$$

$$896 = 21.333333 \times h$$

$$\frac{896}{21.33333} = h$$

$$42 = h$$

Now, just fill in the grid, and you're done.

Unable to use the Reference Sheet, a woman stands at her doorway hoping that someone will give her the formula 2(lw) + 2 (hw) + 2(lh).

The Big Five and the Little Three Questions, Math-Style

The Math FCAT emphasizes certain types of problems over others. Learning about these categories will help focus you as you prepare for the test. The categories have been broken down into the Big Five (multiple-choice question types) and the Little Three (recurring performance task themes)—for your easy learning pleasure. Of course, there are bound to be some questions that don't fit into these categories. When that is the case, say the word *Nepali* to yourself three times in a row, and then look at the problem again. You might find a way to attack it, but if you can't, use Process of Elimination to rule out obvious wrong answers and then guess. But whatever you do, don't leave it blank!

Imagine that on the Math test, Process of Elimination was like a group of animals with two dominant species:

> **Species 1**: *Numeralis eliminatorius*
> This species eliminates answer choices that contain the same numbers used in the original question.
>
> **Species 2**: *Ballparkius nonsensicalus*
> This species eliminates answer choices that contain numbers that are obviously too high or low.

Both POE species can be found throughout the Math FCAT, and in shaded, woodland areas. These animals feed off of tubers, roots, and students who rush through the multiple-choice section of the test. Species 1: *Numeralis eliminatorius* is especially fond of catching students, who, when confused by a question, pick the answer choice that has identical numbers to those in the problem. If captured and trained, both POE species make excellent test-taking techniques.

Here's an example of *Numeralis eliminatorius* in its native habitat.

3. Prakash is assembling a crib for his nephew. He tries a $\frac{3}{8}$-inch screw in one of the legs, but discovers that it is slightly too large. Which of the following screw sizes, in inches, is the next smaller one?

   A. $\frac{1}{8}$

   B. $\frac{5}{16}$

   C. $\frac{1}{2}$

   D. $\frac{5}{8}$

Here it would be easy for you to become prey to *Numeralis eliminatorius*: You look at question 3 and try to solve it in your head, writing nothing down. Nothing comes to mind, so with a quick glance over the answer choices you see choices A, $\frac{1}{8}$, and D, $\frac{5}{8}$, both of which contain 8 in the denominator. Since 8 was also in the original problem ($\frac{3}{8}$-inch screw), you pick A or D and then rush off. *You have just been eaten by Numeralis!*

To solve question 3 mathematically, find a common denominator (16) for all the fractions; so $\frac{3}{8}$ becomes $\frac{6}{16}$, $\frac{1}{8}$ becomes $\frac{2}{16}$, $\frac{5}{16}$ stays unchanged at $\frac{5}{16}$, etcetera. You would see that choice B is just slightly smaller than the $\frac{6}{16}$-inch (or $\frac{3}{8}$-inch) screw.

If you had no idea how to solve question 3, your best bet would be to rule out choices A and D and then take a guess.

Strategy

Be wary of any answer choices that contain the same numbers as those in the question itself. They're often traps designed to trip you up if you hurry.

The other species of POE, *Ballparkius nonsensicalus,* can be found in the problem below.

4. A hat contains 90 slips of paper that are either blue, red, or yellow. If half the slips are yellow, and the ratio of blue slips to red slips is 2 to 1, how many red slips are there?
   A. 60
   B. 45
   C. 30
   D. 15

Common sense can help uncover the incorrect *Ballparkius* answers. First, there are a few things you can ask yourself to determine the general ballpark of the answer. If half the slips are yellow, then how many yellow slips are there? The answer is half of 90, which is 45. That leaves 45 slips of either red or blue paper. With this in mind, do you think the number of red slips is greater or less than 45?

Now, common sense would tell you that there have to be fewer than 45 red slips of paper. This means choices A, 60 and B, 45, are both wrong. A is way off, since it's greater than 45. By estimating the correct answer, you used POE to eliminate *Ballparkius nonsensicalus* answer choices.

Look at the two remaining answer choices, C, 30 and D, 15. The question states, "the ratio of blue slips to red slips is 2 to 1." So which one would you pick? If you choose D, 15, you'd be correct—since 30 blue slips outnumber 15 red slips by 2 to 1.

Strategy

When working multiple-choice problems on the Math FCAT, remember that POE is your most effective test-taking tool.

Sadly, POE is not effective on the gridded-response questions, which is why you should wait and do them only after you've finished the multiple-choice questions. Even so, many gridded-response questions do fall into one of the following categories, and there are other strategies that can be applied.

The Big Five Multiple-Choice Question Types

1. Number Sense Questions

These questions test your understanding of a variety of basic math terms. Favorite topics include: percentages, number lines, mean and median, fractions, ratios, exponents, order of operations, greater than/less than, positive/negative numbers, and number lines. If you just said, "Huh?" to any of those topics, your first order of business is to review them intimately.

Strategy

Questions involving simple math terms do in fact become "simple" if you're familiar with the terms being used.

5. Listed below is the distribution of students at Cloverleaf High who are taking a foreign language.

French	44%
Spanish	34%
German	20%
Urdu	2%

If there are 550 students taking a foreign language, how many of them would you expect to be taking French?

6. $(2 + 3 \times \sqrt{25})^2$ equals
   A. 100
   B. 289
   C. 625
   D. 900

Here's a sneaky question: Did you work on question 5 before doing Question 6? If you did, you spent time on a gridded-response question when there was a simpler multiple-choice question after it. Granted, this book isn't the test, but good habits learned here will transfer automatically to the test itself.

As for question 6, this question tests your knowledge of the proper *order of operations.* Just plugging the numbers into calculator from right to left will yield 625, but this in incorrect. In fact, the *25* in the question indicates that this is a bit of *Numeralis eliminatorius.* Take the square root of 25, multiply it by 3, then add 2, and finally square that number to get choice B, 289.

Having completed the multiple-choice question, it's on to question 5, a fairly straightforward percentage question. If you're familiar enough with percentages, you'll multiply the number of total students (550) by the percentage taking French (44%, or 0.44). This gives you 242.

Knowing your basic math terms is the key to solving Number Sense questions.

2. Chart/Graph Questions

The Math FCAT is filled with questions involving graphs and charts—it could well be the most common question type you see. The easiest of these merely ask you to interpret some data. Harder ones involve reading the data and then doing something with that information. The hardest questions ask you to choose the graph that best represents a particular situation.

7. The graph shows the amount of money earned by Anju and Jonas while working at a pizzeria.

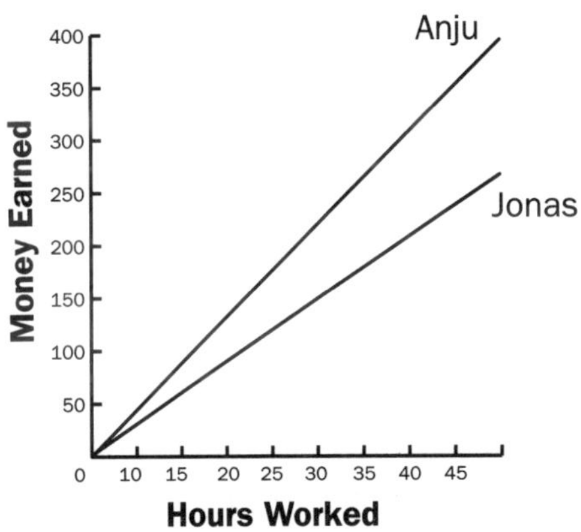

Based on the graph, which of the following is true?

A. Jonas works more hours than Anju each week.

B. Jonas makes more money per hour then Anju does.

C. Anju could work fewer hours and still earn as much money as Jonas each week.

D. Even if Jonas worked the same numbers of hours as Anju, he can never make the same amount of money she does.

8. If Anju and Jonas both work 20 hours one week, what is the best estimate of the mean amount of money earned by them?
   A. $300
   B. $200
   C. $150
   D. $100

9. If management decides that Anju will be paid a flat rate of $400 a week, and that Jonas will receive a raise of $1 per hour, then which of the following statements is true?
   A. The slope of Anju's line will be horizontal.
   B. The slope of Jonas's line will decrease.
   C. The slopes of both Jonas and Anju's lines will increase.
   D. The slope of Anju's line will be vertical.

While you probably won't see three questions based on the same graph, you might see two consecutive questions. In any event, questions 7–9 illustrate three common graph questions. In question 7, you're being asked simply to read the graph correctly. If you do, you'd narrow it down to C or D. Give C a good, hard look, as *could* is a nice word. C is the right answer: D is incorrect if both Anju and Jonas work 0 hours each week.

In question 8, reading the graph is still important, but now you must combine it with number sense terminology, *mean*. *Mean* stands for *average*, and the average of Anju and Jonas's earnings will be the point midway between their two earnings line, straight above 20 hours. Reading across from that point gets you approximately $100, choice D.

In question 9, you have to understand the graph well enough to realize that if Anju receives a flat rate, her payment line won't change in relation to her hours worked. That means her payment line would be a horizontal line coming out of $400. Choice A is the answer.

In order to do well on chart/graph questions, you need to be comfortable interpreting data in a visual form. A good place to find graphs is in the newspaper, especially the business section. Or you can surf the Internet, as many sites undoubtedly use graphs for information. If you want to move on to the bonus level, construct your own graphs. Graph the number of canned goods in your kitchen, or the number of times the phone rings each week. Go to the nearby mall and make a chart showing how many people gave you a dollar when you asked them politely, or the number of people who said yes when you asked them to the prom. With those last two, you'll not only learn about graphs, but you'll learn about love and money as well.

> Find at least ten graphs and analyze them thoroughly; that means pie graphs, bar graphs, and line graphs. Make sure you understand what the horizontal and vertical axes represent.

Not learning all you can about graphs and charts on the Math FCAT is a criminal offense.

3. Problem-Solving Questions

Problem-solving questions are usually word problems that require you to use variables to determine an answer. The plus side is that they're often easy to recognize; just look for variables in the problem and/or the answer choices. The downside is that POE isn't very effective with these questions, since, first, you can't ballpark an answer with variables, and second, all of the answer choices will have numbers from the original question. Still, knowing that POE doesn't work is important, because you'll know not to use it.

Strategy

Problem-solving questions are often recognized by the presence of variables. The first thing you should do to solve them is to label what each variable represents.

10. Bernie's Discount Emporium sells four types of lawn chair. The small version is \$45, the large version retails for \$74, and the deluxe chair is \$145. The store sold \$$t$ worth of lawn chairs in January. If in January, s represents the number of small lawn chairs sold, l represents the number of large lawn chairs sold, and d the number of deluxe chairs sold, which algebraic equation below represents the money received in January on lawn chair sales?
    A. $45s + 74l - 145d = t$
    B. $74s + 45l + 145d = t$
    C. $45d + 74l = t + 145s$
    D. $t = 74l + 145d + 45s$

Lots of writing, lots of variables = problem-solving question. There are four variables in question 10, and if you keep track of what each one represents, you should be able to find the right answer. Going one variable at a time, *s* stands for the small chairs worth \$45, so we can cross out any answer choice that doesn't combine *s* with 45 (the price of small lawn chairs). That gets rids of B and C, so now it's a fifty-fifty chance between A and D. In both of those choices, each of the variables *s*, *l*, and *d* is properly aligned with its respective price, but to find the total money received on lawn chair sales, you would add all three variables. In A, the deluxe lawn chair is subtracted, so it's incorrect. That leaves D, your answer.

Go one step at a time, don't get flustered by the extensive writing, keep track of your variables, and problem-solving questions will be at your mercy.

4. Geometry Questions

Along with graph/chart questions, geometry questions are among the most prevalent on the Math FCAT. Of course, geometry is a broad category that covers a lot of information, but if you want to narrow your focus a bit, then look at the table below:

FCAT's Favorite Geometry Categories:

- measure of angles
- perimeter and area
- questions involving circles and degrees
- triangles, triangles, triangles

Occasionally you'll have a geometry question that you can solve just by looking at it, but these aren't as numerous as they once were. Still, if you're provided with a figure to work with, always take advantage of this by ballparking the correct answer.

11. When the clock is at 4:44, what is the angle between the minute hand and the hour hand?

A. 168 degrees

B. 122 degrees

C. 90 degrees

D. 60 degrees

Look at the hands of the clock and ask yourself, "Is that angle *greater or less* than 90 degrees?" A right angle is 90 degrees, and since the hands are clearly set at an angle more than 90 degrees, you can cross out C and D. They're too small. Now, take a guess, or do the math. B is the answer.

Strategy

On geometry questions that have diagrams, try using your eyes to ballpark the correct answer. This can work with both angles and lengths.

For the most part, though, geometry questions on the grade 10 FCAT are a little more esoteric.

12. A triangle has two sides that are 7 cm and 9 cm each. Which of the following cannot be the third side?

A. 2 cm

B. 4 cm

C. 7 cm

D. 14 cm

Questions like this are why you'll want to know everything there is to know about triangles before you take the Math FCAT. For question 12, use POE first: What can you eliminate? If you said C, then good for you! It's that pesky *Numeralis eliminatorius* again.

Barring a math epiphany, attack question 12 by attempting to sketch the answer choices freehand—but with the right proportions. Draw a triangle with sides of 7 cm, 9 cm, and 2 cm (choice A). Can you do it (without bending time and space)? Probably not, since you need more than 2 cm to close up/connect 7 cm and 9 cm lines and still have a triangle. Since 2 cm cannot be the third side, A is your answer.

Strategy

Know all that you can about the FCAT's favorite geometry categories, and be prepared to use that information.

5. Probability Questions

1. What are the odds that you'll encounter at least one probability question on the Math FCAT?
   A. Pretty good
   B. About 10 in 25 or 6 to 4
   C. No chance at all
   D. What are "odds"?

Hopefully, you picked A, pretty good. Probability questions come in two types:

Type 1: either you figure out all the different possible combinations, or

Type 2: you figure out the odds of an event's occurring.

In type 1, the trick is to think multiplication.

13. The following action figures are sold at a nearby store:

Soldiers	Dinosaurs	Villains	Superheroes
Capt. Bob	Stegosaurus	Dr. Footlock	Mr. Goodness
Audie	Sleestack	Goobertor	Ariadne Jones
Sgt. Mekmek	Cerotopus	Skimjeevil	
	Dimetrodon		

If you went to the store and bought one of each type of action figure, how many different combinations could be made?

A. 72

B. 36

C. 12

D. 4

The phrase *how many different combinations*, or words that mean the same thing, indicate that you're dealing with the first type of probability question. Finding an answer for this type of question always involves multiplication. To solve question 13, list all the different types of action figure, and then multiply them all together.

3 soldiers x 4 dinosaurs x 3 villains x 2 superheroes =
3 x 4 x 3 x 2 = 72, answer A.

For the other type of probability question, the *what are the odds* type, you'll need to find two key numbers: the total number of outcomes, and the number of specific outcomes in which a specific event/item will occur. Look at the next question, that's similar to the earlier question 4.

14. A hat contains 90 slips of paper that are either blue, red, or yellow. Half of the slips are yellow, and the ratio of blue slips to red slips is 2 to 1. If one slip is chosen at random, what is the probability that it is red?

    A. $\frac{1}{2}$

    B. $\frac{1}{3}$

    C. $\frac{1}{6}$

    D. $\frac{1}{12}$

The first number we need to find is the total number of outcomes, which we're given as 90, the total number of slips. The second number we need to identify is the number of desired outcomes in question, the number of red slips. To do this, we want to first find how many yellow slips there are. We know that half the slips are yellow (45 slips), so that means there are 45 slips remaining. Since the ratio of blue slips to red slips is 2 to 1, this means there are 30 blue and 15 red slips. So 15 is our second number, which we must now place over our first number, 90. This gives you $\frac{15}{90}$, or reduced, $\frac{1}{6}$, choice C.

Strategy

On "how many different combinations" questions, think *multiplication*. On "what are the odds" questions, find the two numbers needed to represent the probability (*number of desired outcomes* and *number of outcomes).*

Most of the multiple-choice questions you'll see on the test will fall into one of the Big-Five Question Types, but be prepared to improvise. Invariably some questions will be unique—a question might combine geometry with fractions, for instance—and while they won't be *exactly* like the ones discussed here, you'll find that the same test-taking strategies can be applied.

Once you've answered all the multiple-choice questions, it's time for the gridded-response problems.

The Little Three Performance Task Question Types

Take a multiple-choice problem of average difficulty, multiply that by *pi*, subtract all the answer choices, bake for thirty minutes at 350 degrees, and you have your typical performance task.

Performance tasks on the Math FCAT at first might seem quite difficult, since they're usually multistep problems. However, the fact that they require more than one step is actually good news for you, as your chances of garnering partial credit increase. You might not understand what part c is asking for, but you have no problems with parts a and b. If that's the case, then answer a and b as completely as possible, give c your best shot, and then move on to the next question. Remember, perfection is not a requirement on the FCAT.

Quick Tips for Open-Ended Performance Tasks

1. Don't spend too much time on any one question.
2. Be prepared for multistep questions.
3. Write down all your work.
4. Warm up your calculator.
5. Be prepared to draw.

While you should be familiar with most of these points already, point 5 could use some additional explanation. You might be asking what you'll need to be prepared to draw. A self-portrait? Dogs playing poker? Instead, the answer is that you'll most likely have to create a chart or graph.

1. Make a Chart/Graph Performance Tasks

Earlier in this chapter, you were encouraged to practice your skill at making graphs. That wasn't intended as just busy work; it was preparation for the performance task section, where you'll likely have to draw at least one graph or chart.

15. A local health club is offering two types of membership. Membership A, which requires no initial fee, has monthly dues of $36 a month. Membership B requires an initial fee of $84, but after that monthly dues are $24.
    a. Suppose you wish to use the gym for eight months. What would be the total cost if you bought Membership A? What would be the total cost if you bought Membership B?
    b. Over a period of one year starting in January, plot a graph showing the cost of each membership on a month-to-month basis.
    c. Compare the two memberships and explain which offer is the most economical.

Part a of this question requires use of your calculator, Part b requires your graphing skills, and part c . . . well, let's worry about that later. Don't forget to write down all your work.

Part a: Let's establish an equation that represents both memberships.

Membership A = 8 months × $36 = $288
Membership B = 8 months × $24 + $84 (initial fee) = $276

We've just earned 1 point right there. Membership B is cheaper after eight months, but just barely.

Part b: First, label the graph with the appropriate axes, which is always the first thing you should do when making a graph. Let's put the months along the horizontal axis, and the dollar amounts along the vertical axis.

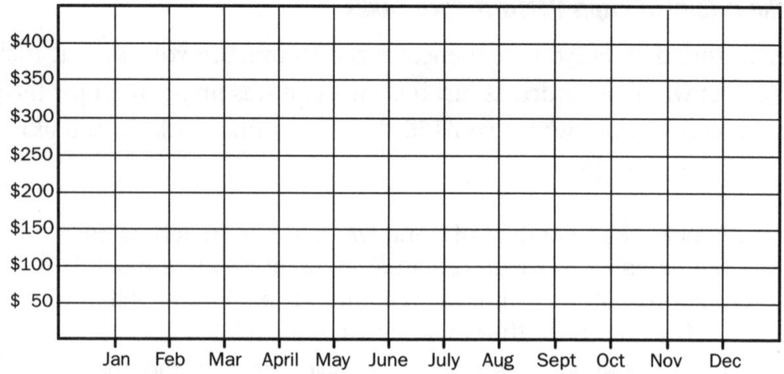

Graphing both membership types can be done fairly simply, since we already know two points. For membership A, in January, the cost is $36, and in August, it's $288, since that's the eighth month. Therefore, if you draw a line between (January, $36) and (August, $288), you'll have a line graph that represents Membership A's cost.

The same can be done for Membership B, although in this case, the two points are (January, $108) and (August, $276). January is higher for Membership B because of the initial fee that's charged.

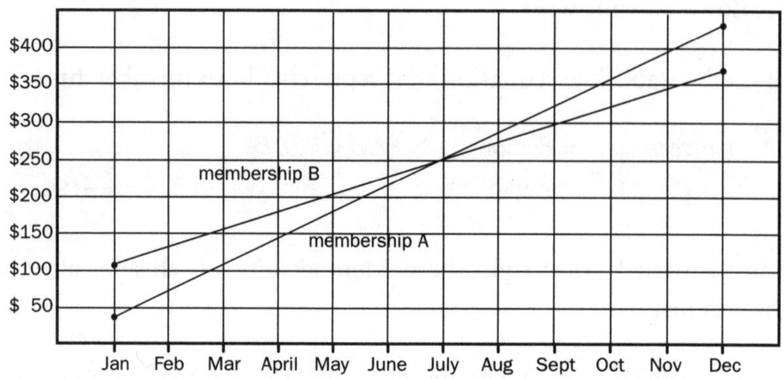

Part c: Now that the graph is complete, this part isn't as difficult. By looking at the graph, you can see that Membership A is more economical than B in the first six months. In the seventh month, July, the costs of the two memberships are equal ($252) but after that, Membership B is a better choice.

Simply writing that sentence and referring to the graph should be all you need to complete part c. Hooray!

Strategy

Get very comfortable with making graphs of all types: pie graphs, bar graphs, line graphs. Learn them, love them, live them!

A student nervously prepares to complete the bar graph entitled "Number of Angry Lizards in Cages 1, 2, and 3."

2. Two-Dimensional Geometry Performance Tasks

3. Three-Dimensional Geometry Performance Tasks

If you read the titles above and got the feeling you're going to see some performance tasks that ask about geometry, you've hit the nail (thin metal cylinder with a cone at the end) right on the head (circle). Although these open-ended geometry questions can't be as easily pinned down as their multiple-choice counterparts, there are some things you can do to prepare for them.

Strategy

Know the formulas for area and perimeter for the following two-dimensional figures:

- Triangle
- Circle
- Square
- Rectangle

Know the formulas for volume for the following three-dimensional figures:

- Pyramid
- Sphere, cylinder, cone
- Cube
- Rectangular solid (a box shape)

The formulas for some of these figures can be found on the Reference Sheet in your FCAT test, but not all, so don't take any chances. Finding the area or volume of a figure might be the first step on one of these open-response problems, and determining the correct area or volume will make the rest of the problem much easier.

Beyond this fact, the geometry questions are not very predictable. If you keep in mind the quick tips about open-response questions, you should survive unscathed. However, if you do feel like you're getting in way too deep on a question, just shout "ma*n* dubáu*n*nu!" which means "I'm drowning!" in Nepalese. A lifeguard will come by shortly to rescue you.

Chapter 4

WRITING

HOW THE FCAT WRITING PROMPT IS LIKE A STICK OF BUBBLE GUM

By itself, a stick of bubble gum isn't very interesting. It's nothing more than a flat, strangely colored rectangle made out of a substance that looks a lot like cardboard. But once you place the gum in your mouth, everything starts to change. First, you get a burst of flavor, and as your jaw muscles start flexing, the gum takes on an ever-changing variety of shapes. Place it between your lips, exhale, and the flat stick of gum has transformed into a bubble-gum bubble.

Similarly, the FCAT writing prompt is nothing more than a string of words. It's up to you to take the prompt and use it to create a concise, insightful essay that will earn you a high score on this part of the test. To do this you'll have to read the prompt, flex your mental muscles to come up with the key ingredients for your composition, and then exhale those words onto paper. The end result should be an essay that is as contained and logical as a sphere of chewing gum.

Although you're probably well-versed in the basics of essay writing, this chapter will review what's expected of you on the Writing FCAT.

A solitary student walks the moors preparing for the FCAT essay.

Writing FCAT: Format

Quite simply, the Writing FCAT gives you 45 minutes to write an essay on a specified topic. On the day of the test, the schedule should run as follows:

1. Students arrive and sit down in desks.
2. Students plead with administrator in charge to let them not take the test.
3. Administrator either
   a) says "no"; or
   b) does not dignify statement with a response.
4. Students receive a writing prompt (assignment/instructions). It will be either a persuasive or an expository prompt.

5. Students now have 45 minutes in which to write their essay. Only the completed essay, and not any notes, will receive a grade of 0–6 points. Students will receive a separate sheet for planning and organizing their thoughts but must submit the final essay on the two lined pages provided.

The Writing FCAT, originally known as the *Florida Writes!* test, is now a requirement for graduation. The FDOE has considered expanding the test to include multiple-choice questions, but the test remains unchanged as of this printing.

Scoring

The methods used to score your essay provide interesting insight on how you should approach this part of the test. Your essay will be graded by two readers, each of whom will assign it a score from 1 to 6 (highest). You'll need to get at least a score of 3 to pass. If the graders disagree on a score, say one gives your essay a 4, while the other assigns it a 3, then your score will be a 3.5. On differences greater than 1 point, a third reader may be brought in.

The readers (primarily Florida teachers) make a judgment about the entire essay without focusing on any one aspect of writing. However, they base your score on the following objectives:

Writing FCAT Objective	What it Means
1. *Focus*	How well does the essay present and maintain a clear theme or idea?
2. *Organization*	Is there a coherent structure to the development of the essay, such as a beginning, middle, and end? Are transitional devices used properly? Are the sentences and ideas connected to each other? Is there a concluding statement?

3. *Support* — What is the quality of the details used to support the main idea? Are the details credible, thorough, and elaborate? Are there examples and illustrations that clearly support the topic?

4. *Conventions* — Does the essay have proper punctuation, spelling, capitalization, and variation in sentence structure?

Hypothetically, then, if you write a deeply moving, well-developed essay with atrocious grammar, you might still get a 3, which is the minimum passing score. This is because three out of the four objectives have to do with *content*, not *grammar*. On the other hand, if you write an absolutely bland essay with perfect grammar, you might not get 3 points, since your essay lacks focus, organization, and support. With this in mind, focus more on developing content persuasively and concisely rather than on using perfect grammar.

Strategy

When planning your essay, concentrate on the content and don't worry as much about the grammar.

You can always go through your essay before time runs out and check over your spelling and grammar, correcting any mistakes that you may find. However, if you write an unconvincing essay from the start, you're sure to be on the path to a lower score.

While content bears more weight here than grammar, your essay still has to be comprehensible. A sentence so jumbled that it makes no sense won't help you at all, since the reader won't be able to understand the point you are trying to make. Consider the following three sentences.

A. I believe students should be allowed to vote in national elections.
B. I beleive the students they should be aloud to vote national elections.
C. I believe the students they vote and national elections.

Version A has good grammar and clear content. Version B has some misspellings and incorrect grammar, but the writer still makes his point. Version C, however, is grammatically incorrect and unclear. Shoot for version A, but if you have to settle for version B, bear in mind that sentence B is better than sentence C.

Keep your ideas simple and clear. Complicated ideas are harder to explain, and the graders don't expect anything fancy or unusual. They won't be impressed by a dazzling subject if your argument isn't clearly made. You're better off making a solid, well-developed essay, even if it isn't on the most awe-inspiring subject. Having said that, you do want to make your essay as compelling as possible, so try to include examples and supporting details that will engage the reader.

Writing Prompts

On the test, you'll see either an *expository* or a *persuasive* prompt. An expository prompt asks you to explain a topic or describe a concept, while a persuasive prompt asks you to convince the reader of a certain point of view. Since the two prompts ask you to do two different things, you should first determine what kind of essay you're being asked to write.

Strategy

First, figure out whether you're being asked to write an expository or a persuasive essay. In other words, are you being asked to *explain* something or to *persuade* the reader of something?

This isn't that hard to do. Expository prompts often have the word "explain" in them, while persuasive prompts often contain the phrase "your point of view." Both essays require you to use specific examples and information to make your essay highly detailed and very persuasive.

Examples of Expository Prompts:

- Think of a job you would like to do. Now explain why you would like to do that job.
- Think of all the places in the world. Now explain why one place is your favorite place.

Examples of Persuasive Prompts:

- Write a letter to convince your parents to accept your point of view on the effects watching TV has on grades.
- Many adults think that having a curfew helps prevents crime. Write an essay on whether or not you believe a curfew is an effective deterrent to crime.

Time to Flex your Mental Muscles

After reading the prompt, you should have around 43.7692 minutes left in the planning session. That's a lot of time, so don't feel pressured to start writing immediately.

Strategy

Give yourself about 10–15 minutes to brainstorm as many thoughts as possible. Write them all down.

The key to good brainstorming is to make some conclusions on your own, using the text to back up your ideas. To illustrate this idea, look at this writing prompt again:

- Many adults think that having a curfew helps prevents crime. Write an essay on whether or not you believe a curfew is an effective deterrent to crime.

First, you have to decide what position you want to take on this subject. It doesn't matter what side of the argument you take; all that matters is how *well* you support your argument. There's no correct answer, only a well-supported argument versus a poorly supported argument.

Strategy

For a persuasive prompt, it's not important which side of the argument you take, only that you support your argument thoroughly and concisely.

Let's say you're going to write a persuasive essay which states that curfews are an effective deterrent to crime. Take some time right now to brainstorm as many ideas as possible that will support this statement. Do this on your own, and then compare it to the list below:

Do not plan your essay on the two lined pages provided in the test booklet, since that space is needed for the essay and nothing else. Use the separate sheet provided to jot down notes instead.

Pro-Curfew Persuasive Essay: Brainstorm Ideas

1. Most crimes occur at night; having a curfew will limit a criminal's ability to get around.
2. Random crimes of opportunity, like breaking a car window to get to some CD's, should decrease.
3. Police would have an easier time preventing crime.
4. Infringement on people's freedom not total—still a lot of time to roam streets.
5. A one-year study could be done to see if curfew is an effective deterrent.
6. Human-eating robots come out most often a night; curfew will help prevent crimes of digestion.

As you can see from item 6, some ideas are not as good as others. Don't sweat it: the whole point of the brainstorming session is to come up with as many ideas as possible, even though you probably won't use all of them. If an idea doesn't fit well into the development of your essay, don't feel obligated to use it. It's better to elaborate on a few good ideas than to just list a whole bunch of items that are weakly explored.

After you finish brainstorming, take some time to organize your thoughts. Use the planning sheet to develop your argument before you write the whole thing out. Try writing out an informal outline to ensure that you have included all the major elements. Decide which ideas you'll use, what order they should go in, and how you'll transition all of them. Transitions are important in essay development; it's up to *you*, not the reader, to link your ideas together. Are these ideas similar, and if so, how? Do they conflict? How?

The goal of the organization stage is to get a sense of what your final essay will look like. Your yet-to-be-formed essay should have a main idea (stated in the opening paragraph), a body of supporting details, and a conclusion.

To return to the bubble gum analogy, if brainstorming is where you chew the gum, then organizing the essay is where you line up the gum between your lips in order to make the bubble. It's an important stage: If you place the gum in the wrong place, your bubble will pop before it ever gets big. If you don't organize your thoughts well, your essay will never reach the level of detail and clarity needed to get a high score.

Details, Details, Details

Precise, elaborate details are crucial to any essay. To illustrate their importance, read the following sentences, which range from *vague* to *elaborately detailed.* As you can see, providing sufficient details can mean the difference between staying alive and becoming tiger chow.

1. There is a cat.
2. There is a large cat over there.
3. There is a large cat coming toward you.
4. There is a tiger running at you.
5. There is a vicious, five-hundred pound tiger running rapidly toward you; he just escaped from the circus, and boy, is he tired of having chairs stuck in his face.

Luckily, the Writing FCAT is not so high-stakes that you have to write a perfectly composed and compelling essay. However, you should make it a point to elaborate all of your brainstormed ideas, clarifying their meanings as well as possible. For instance, look at the following statement.

> Random crimes of opportunity—like breaking a car window to get to some CDs—should decrease.

The statement above is far too incomplete. To correct this, add something like this:

> With a curfew, many random crimes of opportunity should decrease. Many times teenagers, who commit a fair amount of vandalism and petty theft, are simply bored.

They walk around the neighborhood looking for something to do, and often end up doing something foolish and criminal —like breaking a car window to get to some CDs—just for a stupid thrill. However, with a curfew in place, the opportunity to commit random crimes would decrease, since teenagers would not be as likely to go out, and if they do, they would have to be much more cautious in their activities. While an increase in bored teenagers indoors might lead to an increase in prank phone calls, at least property damage should decrease. And, these days, there's always caller identification!

The response above takes the original idea and elaborates on its meaning in a very precise manner. The more of this you can do, the more persuasive your argument will be, giving your essay score a boost.

The Opposing Argument

When writing a persuasive essay, don't forget that there's always another valid viewpoint on the subject. The fact that you chose the procurfew side doesn't mean that the anti-curfew side is completely wrong. So put away your procurfew posters, cancel the procurfew rally you planned, and take some time to think about what reasons would validate the anticurfew argument. Include one or two of those in your own essay, and follow them up with your own explanation as to why you disagree.

> The essay skills you use on the Writing FCAT will also help you with the 2- and 4-point questions on the Reading FCAT. These questions are often just mini-essays.

Strategy

Try to present an understanding of *both sides* of an argument. If you can describe the other side's position, and reasons why you disagree with it, then you have strengthened your essay.

Looking back at the list of brainstorming ideas, you'll see that in items 4 and 5, there's an attempt to see the anticurfew side of things, and a subsequent attempt to counter those arguments. That portion of your essay could look something like this:

> There are many people who will argue that a curfew is unconstitutional because it infringes on their freedom. While this is a valid concern, there are steps that could be taken to limit this argument. First, the curfew should be late enough that the majority of the population wouldn't even notice, or mind its existence—say, from midnight to 6:00 A.M. While this curfew would prevent insomniacs from taking a stroll at 3:00 A.M., it would also help protect the community, and as a whole everyone would benefit. Besides, the insomniacs in question would have 18 other hours in the day to walk, not an unreasonable amount of time. Second, the curfew could be a temporary one, perhaps for only one year, to judge whether it is an effective deterrent. If it isn't effective, or if it is found to deter only a small percentage of crime, then it could be abandoned. But if the curfew leads to a significant decrease in criminal activity, which is very possible, then most people will probably want to keep it in place.

The paragraph above acknowledges the opposition's argument and counters it. While you shouldn't spend the entire essay attacking your opponent's position, one paragraph on the subject is an excellent idea.

Hey Mumbles! Be Sure to Write Neatly

If your cursive handwriting is so bad that sometimes even *you* don't know what you've written, make sure you print out your essay. An illegible essay cannot be graded, and it would be a tragedy if a well-written essay lost points simply because the graders—who have to read a lot of essays, mind you—could not read what you have written. Even a partially illegible essay may suffer at score time as well.

Pacing Suggestions: Essay

Brainstorm ideas	**12–15 minutes**
Organize	**5 minutes**
Write essay	**20 minutes**
Proofread	**3–5 minutes**

If you can plan, write, and proofread a tidy three- to five-paragraph essay in 30 minutes, don't force yourself to use the pacing suggestions in the sidebar. There's no need to pad an adequate essay. Nor should you just ramble on until time is up; that will muddle your argument and even increase the likelihood you'll make spelling, punctuation, and grammatical errors. If you're a speedy writer, use any extra time to check your work.

After completing your final draft, you should have about 3–5 minutes left in the session. Use this time to proofread for mistakes or to edit/rewrite any very awkward sentences. Also, make sure each paragraph leads smoothly into the next. If you make any corrections, do so neatly.

Strategy

Make sure you proofread your essay and correct any errors in grammar, punctuation, and sentence construction.

Once you have finished checking your essay for errors, you're done! Now that the Writing FCAT is under your belt, you have several weeks off before the rest of the FCAT tests are given. There's nothing to do but kick back, relax, and maybe . . . chew some gum.

Chapter 5

RESULTS

HOW YOUR GRADE 10 FCAT TEST SCORES ARE EXACTLY LIKE YOUR GRADE 8 FCAT TEST SCORES

At last, a chapter title that speaks for itself. If you remember how the FCAT is scored from your grade 8 results, then you'll already know this information. But if you wiped all that information from your mind right after you took that test, you might want this refresher.

Part of your test is hand scored by teachers/evaluators, and part of it is machine scored. Your scores will be broken down into three main categories:

1. A scaled score
2. An achievement level score
3. A percentage subscore

The *scaled score* is a number between 100 (lowest) and 500 (highest). The test is "scaled" so that the average score in the state is around 300. So if you receive a 340, you have scored better than average.

You'll also receive a National Percentile Rank for Math and Reading. This score relates only to the March tests, not the February tests. Since the FCAT tests you take in February are what determine whether you may graduate, it is these scores—and not the National Percentile Rank—that have more direct bearing on you.

Average Statewide Score Results

	1998	1999
Grade 10 Reading	299	306
Grade 10 Math	302	312

Your scaled score is then converted into an a*chievement level* score from 1 to 5 (highest). In 1999, the scaled-score range for a Level 4 math score was 340–374, so a scaled score of 375 or higher got a Level 5, and a scaled score of 339 earned a Level 3. The achievement levels are described as follows:

Achievement Level	Description
5	You were successful with even the most challenging content in the Sunshine State Standards.
4	You were successful with the state standards.
3	You had partial success, but your performance was inconsistent. This is the lowest passing achievement level.
2	You had only limited success with state standards.
1	You had little success with state standards.

Achievement level 5 is very difficult to achieve; in 1999, very few students (5 percent in reading and 4 percent in math) received a level 5. A plan is currently in place to make the test scoring even harder in the future.

Grade 10 FCAT Math Achievement Levels					
	Level 1	**Level 2**	**Level 3**	**Level 4**	**Level 5**
1999–2001	100–286	287–314	315–339	340–374	375–500
2002–up	100–299	300–327	328–352	353–387	388–500

Grade 10 FCAT Reading Achievement Levels					
	Level 1	**Level 2**	**Level 3**	**Level 4**	**Level 5**
1999–2001	100–286	287–326	327–354	355–371	372–500
2002–up	100–299	300–339	340–367	368–384	385–500

So for example, if you score a 330 on the Grade 10 Reading FCAT in 2001, that would put you at Level 3. But if you get a 330 on the same test in 2002, you'll be scored at Level 2.

In addition to the scaled score and the achievement level, you will also receive a *percentage score* in the reading and math content categories. The percentage subscores show how well you performed in various subcategories, such as geometry, algebraic thinking, and so forth. So a math subscore of 75 percent for the geometry strand means that you have achieved 75 percent of the geometry strand content. These subscores are interesting from a teaching viewpoint, as they can show your teacher what subjects you are strongest and weakest in, but as far as your graduation is concerned, they are not relevant.

While a low achievement-level score can be a cause for concern, it isn't necessarily an indication that you are lagging far behind in your studies or that your efforts have been worthless. It's important that you keep these scores in perspective. If you don't pass an FCAT subject test the first time you take it, it's understandable you'll be disappointed. But the fact of the matter is that only about half of all Florida tenth graders have passed the Math test each year, and only about one-third have survived the Reading test. So if you pass all the FCAT tests, consider yourself fortunate, and if you have to take one or more test over again, you should realize that many of your peers will be in the same situation as you.

In terms of how the FCAT is scored, this is the end of the story. But here, the *end of the story* only means *end of the discussion on how you scored on one standardized test.* Consider this test an interesting "checkpoint" along a very long highway. Some students who score at the lowest level on this test will go on to graduate from prestigious universities with advanced degrees, while other students who score at the top will struggle to finish high school. Your scores simply highlight the areas in which you need improvement. The FCAT is a standardized test you need to pass graduate high school, but it says nothing about you as a person. So regardless of how you perform on it, don't ever doubt your ability to succeed. Of all the advice given in this book, nothing is more important—or more accurate—than this.

How Did We Do? Grade Us.

Thank you for choosing a Kaplan book. Your comments and suggestions are very useful to us. Please answer the following questions to assist us in our continued development of high-quality resources to meet your needs.

The Kaplan book I read was: ______________________________

My name is: ______________________________

My address is: ______________________________

____

My e-mail address is: ______________________________

What overall grade would you give this book? (A) (B) (C) (D) (F)

How relevant was the information to your goals? (A) (B) (C) (D) (F)

How comprehensive was the information in this book? (A) (B) (C) (D) (F)

How accurate was the information in this book? (A) (B) (C) (D) (F)

How easy was the book to use? (A) (B) (C) (D) (F)

How appealing was the book's design? (A) (B) (C) (D) (F)

What were the book's strong points? ______________________________

How could this book be improved? ______________________________

Is there anything that we left out that you wanted to know more about?

Would you recommend this book to others? ☐ YES ☐ NO

Other comments: ______________________________

____

____

Do we have permission to quote you? ☐ YES ☐ NO

Thank you for your help. Please tear out this page and mail it to:

Dave Chipps, Managing Editor
Kaplan Educational Centers
888 Seventh Avenue
New York, NY 10106

Or, you can answer these questions online at www.kaplan.com/talkback.

Thanks!

About KAPLAN

Kaplan, Inc. is one of the nation's leading providers of education and career services. Kaplan is a wholly owned subsidiary of The Washington Post Company.

KAPLAN TEST PREPARATION & ADMISSIONS

Kaplan's nationally recognized test prep courses cover more than 20 standardized tests, including secondary school, college and graduate school entrance exams, as well as foreign language and professional licensing exams. In addition, Kaplan offers a college admissions course, private tutoring, and a variety of free information and services for students applying to college and graduate programs. Kaplan also provides information and guidance on the financial aid process. Students can enroll in online test prep courses and admissions consulting services at www.kaptest.com.

Kaplan K12 Learning Services partners with schools, universities, and teachers to help students succeed, providing customized assessment, education, and professional development programs.

SCORE! EDUCATIONAL CENTERS

SCORE! after-school learning centers help K–10 students build confidence along with academic skills in a motivating, sports-oriented environment.

SCORE! Prep provides in-home, one-on-one tutoring for high school academic subjects and standardized tests.

eSCORE.com is the first educational services Web site to offer parents and kids newborn to age 18 personalized child development and educational resources online.

KAPLANCOLLEGE.COM

KaplanCollege.com, Kaplan's distance learning platform, offers an array of online educational programs for working professionals who want to advance their careers. Learners will find nearly 500 professional development, continuing education, certification, and degree courses and programs in Nursing, Education, Criminal Justice, Real Estate, Legal Professions, Law, Management, General Business, and Computing/Information Technology.

KAPLAN PUBLISHING

Kaplan Publishing produces retail books and software. Kaplan Books, published by Simon & Schuster, include titles in test preparation, admissions, education, career development, and life skills; Kaplan and *Newsweek* jointly publish guides on getting into college, finding the right career, and helping children succeed in school.

KAPLAN PROFESSIONAL

Kaplan Professional provides assessment, training, and certification services for corporate clients and individuals seeking to advance their careers. Member units include:

- Dearborn, a leading supplier of licensing training and continuing education for securities, real estate, and insurance professionals
- Perfect Access/CRN, which delivers software education and consultation for law firms and businesses
- Kaplan Professional Call Center Services, a total provider of services for the call center industry
- Self Test Software, a world leader in exam simulation software and preparation for technical certifications
- Schweser's Study Program/AIAF, which provides preparation services for the CFA examination

KAPLAN INTERNATIONAL PROGRAMS

Kaplan assists international students and professionals in the United States through a series of intensive English language and test preparation programs. These programs are offered at campus-based centers across the United States. Specialized services include housing, placement at top American universities, fellowship management, academic monitoring and reporting, and financial administration.

COMMUNITY OUTREACH

Kaplan provides educational career resources to thousands of financially disadvantaged students annually, working closely with educational institutions, not-for-profit groups, government agencies and grass roots organizations on a variety of national and local support programs. These programs help students and professionals from a variety of backgrounds achieve their educational and career goals.

BRASSRING

BrassRing Inc., the premier business-to-business hiring management and recruitment services company, offers employers a vertically integrated suite of online and offline solutions. BrassRing, created in September 1999, combined Kaplan Career Services, Terra-Starr, Crimson & Brown Associates, thepavement.com, and HireSystems. In March 2000, BrassRing acquired Career Service Inc./Westech. Kaplan is a shareholder in BrassRing, along with Tribune Company, Central Newspapers, and Accel Partners.